SAFETY OF GENETICALLY ENGINEERED FOODS

APPROACHES TO ASSESSING UNINTENDED HEALTH EFFECTS

Committee on Identifying and Assessing Unintended Effects
of Genetically Engineered Foods on Human Health

Board on Life Sciences
Food and Nutrition Board
Board on Agriculture and Natural Resources

INSTITUTE OF MEDICINE *AND*
NATIONAL RESEARCH COUNCIL
OF THE NATIONAL ACADEMIES

THE NATIONAL ACADEMIES PRESS
Washington, D.C.
www.nap.edu

THE NATIONAL ACADEMIES PRESS • 500 Fifth Street, N.W. • Washington, D.C. 20001

NOTICE: The project that is the subject of this report was approved by the Governing Board of the National Research Council, whose members are drawn from the councils of the National Academy of Sciences, the National Academy of Engineering, and the Institute of Medicine. The members of the committee responsible for the report were chosen for their special competences and with regard for appropriate balance.

Support for this project was provided by the U.S. Department of Health and Human Services Food and Drug Administration under contract number 223-93-1025, the U.S. Department of Agriculture under contract number 59-0790-1-183, and the U.S. Environmental Protection Agency under contract number X-82956001. The views presented in this report are those of the Committee on Identifying and Assessing Unintended Effects of Genetically Engineered Foods on Human Health and are not necessarily those of the funding agencies.

International Standard Book Number 0-309-09209-4 (Book)
International Standard Book Number 0-309-53194-2 (PDF)
Library of Congress Control Number: 2004107570

Additional copies of this report are available from the National Academies Press, 500 Fifth Street, N.W., Lockbox 285, Washington, DC 20055; (800) 624-6242 or (202) 334-3313 (in the Washington metropolitan area); Internet, http://www.nap.edu.

For more information about the Institute of Medicine, visit the IOM home page at: **www.iom.edu.**

THE NATIONAL ACADEMIES
Advisers to the Nation on Science, Engineering, and Medicine

The **National Academy of Sciences** is a private, nonprofit, self-perpetuating society of distinguished scholars engaged in scientific and engineering research, dedicated to the furtherance of science and technology and to their use for the general welfare. Upon the authority of the charter granted to it by the Congress in 1863, the Academy has a mandate that requires it to advise the federal government on scientific and technical matters. Dr. Bruce M. Alberts is president of the National Academy of Sciences.

The **National Academy of Engineering** was established in 1964, under the charter of the National Academy of Sciences, as a parallel organization of outstanding engineers. It is autonomous in its administration and in the selection of its members, sharing with the National Academy of Sciences the responsibility for advising the federal government. The National Academy of Engineering also sponsors engineering programs aimed at meeting national needs, encourages education and research, and recognizes the superior achievements of engineers. Dr. Wm. A. Wulf is president of the National Academy of Engineering.

The **Institute of Medicine** was established in 1970 by the National Academy of Sciences to secure the services of eminent members of appropriate professions in the examination of policy matters pertaining to the health of the public. The Institute acts under the responsibility given to the National Academy of Sciences by its congressional charter to be an adviser to the federal government and, upon its own initiative, to identify issues of medical care, research, and education. Dr. Harvey V. Fineberg is president of the Institute of Medicine.

The **National Research Council** was organized by the National Academy of Sciences in 1916 to associate the broad community of science and technology with the Academy's purposes of furthering knowledge and advising the federal government. Functioning in accordance with general policies determined by the Academy, the Council has become the principal operating agency of both the National Academy of Sciences and the National Academy of Engineering in providing services to the government, the public, and the scientific and engineering communities. The Council is administered jointly by both Academies and the Institute of Medicine. Dr. Bruce M. Alberts and Dr. Wm. A. Wulf are chair and vice chair, respectively, of the National Research Council.

www.national-academies.org

Preface

Genetically modified foods and food products derived from genetically engineered organisms are among a number of biotechnological developments intended to improve shelf life, nutritional content, flavor, color, and texture, as well as agronomic and processing characteristics. Although in popular parlance the term *genetically modified* often is used interchangeably with *genetically engineered*, in this report *genetic modification* refers to a range of methods used to alter the genetic composition of a plant or animal, including traditional hybridization and breeding. *Genetic engineering* is one type of genetic modification that involves the intention to introduce a targeted change in a plant, animal or microbial gene sequence to effect a specific result.

While there are a variety of methods for identifying and measuring specific changes that result from genetic engineering, as well as from conventional breeding techniques, such changes are not always easily discernible—particularly when they are unexpected outcomes of the process or when they result from latent expression of the genetic change or accumulated changes in functional effects in the modified organism.

The addition of genetic engineering to the repertoire of methods to genetically modify organisms has increased the number and type of substances that can be intentionally introduced into the food supply, as well as the magnitude of these changes. While these intended changes can be readily evaluated for their safety in food, unintentionally introduced changes in the composition of foods may be more difficult to identify and assess. Whether genetic engineering per se affects the likelihood of unintentionally introducing undesired compositional changes in food is not fully understood. In contrast to adverse health effects that have been associated with some traditional food production methods, similar serious health effects have not been identified as a result of genetic engineering techniques used in food production. This may be because developers of bioengineered organisms perform extensive compositional analyses to determine that each phenotype is desirable and to ensure that unintended changes have not occurred in key components of food.

Improvement in currently available methods for identifying and assessing unintended compositional changes in food could further enhance the ability of product developers and regulators to perform appropriate testing to assure the safety of food. Whether all such analyses are warranted and are the most appropriate methods for discovering unintended changes in food composition that may have human health consequences remains to be determined.

Scientific advances in agricultural biotechnology continue to improve our understanding of plant crops, microorganisms, and food-animal genetics. Nevertheless, the public health system continues to face many questions about the impact of agricultural biotechnology on human health. As a result of these new scientific advances and public concern about the potential for unintended compositional changes in genetically engineered food that might in turn result in unintended health effects, the National Academies convened this committee to explore the similarities and differences between genetic engineering and other genetic modifications, including conventional breeding practices, with respect to the frequency and nature of unintended effects associated with them—in particular with regard to potential changes in the biochemical composition of plant- and animal-derived foods and methods that would be most useful in assessing the occurrences of unintended changes that might affect consumer health.

ACKNOWLEDGMENTS

The Committee on Identifying and Assessing Unintended Effects of Genetically Engineered Foods on Human Health was aided in its challenging tasks by the invaluable contributions of a number of individuals. First and foremost, many thanks are due to the committee members who volunteered countless hours to the research, deliberations, and preparation of the report. Their dedication to this project and to a stringent time-line was commendable and was the foundation of our success.

Many individuals volunteered significant time and effort to address and educate our committee members during the workshops. Additionally, the committee wishes to acknowledge the invaluable contributions of the study staff: Ann Yaktine, senior program officer and study director; Michael Kisielewski, research assistant; and Sybil Boggis, senior project assistant. The committee also acknowledges other staff members who contributed to the development and initial conduct of this study: Jennifer Kuzma, study director until September 2002; Abigail Stack, study director until February 2003; and Tamara Dawes, project assistant until February 2003. This collaborative project benefited from the general guidance of Allison Yates, director emeritus of the Food and Nutrition Board; and Linda Meyers, the Board's current director; Charlotte Kirk Baer, director of the Board on Agriculture and Natural Resources; and Frances Sharples, director of the Board on Life Sciences. The committee also thanks Geraldine Kennedo for logistical arrangements and Craig Hicks for writing assistance and technical editing.

Bettie Sue Masters, *Chair*
Committee on Identifying and Assessing Unintended Effects
of Genetically Engineered Foods on Human Health

Reviewers

This report has been reviewed in draft form by individuals chosen for their diverse perspectives and technical expertise, in accordance with procedures approved by the NRC's Report Review Committee. The purpose of this independent review is to provide candid and critical comments that will assist the institution in making its published report as sound as possible and to ensure that the report meets institutional standards for objectivity, evidence, and responsiveness to the study charge. The review comments and draft manuscript remain confidential to protect the integrity of the deliberative process. We wish to thank the following individuals for their review of this report:

Arthur J. L. Cooper, Burke Medical Research Institute
Neal First, University of Wisconsin
Michael Grusak, Baylor College of Medicine
Harry A. Kuiper, RIKILT-Wageningen University Research Center
Terry Medley, DuPont Agriculture and Nutrition
Ian Munro, CanTox, Inc.
James Murray, University of California, Davis
Marion Nestle, New York University
Nicholas J. Schork, University of California, San Diego
Margaret E. Smith, Cornell University
Mark Westhusin, Texas A&M University
Walter Willett, Harvard University

Although the reviewers listed above have provided many constructive comments and suggestions, they were not asked to endorse the conclusions or recommendations nor did they see the final draft of the report before its release. The review of this report was overseen by Mary Jane Osborn, University of Connecticut Health Center and Michael P. Doyle, University of Maryland, College Park. Appointed by the National Research Council and Institute of Medicine, they were responsible for making certain that an independent examination of this report was carried out in accordance with institutional procedures and that all review comments were carefully considered. Responsibility for the final content of this report rests entirely with the authoring committee and the institution.

Contents

Executive Summary

BACKGROUND FOR THE STUDY

Genetic engineering and other new technologies are among many advances made to traditional breeding practices in plants, animals, and microbes to enhance food quality and increase productivity. Genetic engineering, the targeted manipulation of genetic material, and nontargeted, nontransgenic methods—including chemical mutagenesis and breeding—are components of the entire range of genetic modification methods used to alter the genetic composition of plants, animals, and microorganisms. (For more comprehensive definitions of key terms used throughout this report, please see Appendix A: Glossary.)

In this report, genetic engineering refers only to recombinant deoxyribonucleic acid (rDNA) methods that allow a gene from any species to be inserted and subsequently expressed in a food crop or other food product. Although the process involving rDNA technology is not inherently hazardous, the products of this technology have the potential to be hazardous if inserted genes result in the production of hazardous substances.

Nongenetic engineering methods of genetic modification include embryo rescue, where plant or animal embryos produced from interspecies gene transfer, or crossing, are placed in a tissue culture environment to complete development. Other methods include somatic hybridization, in which the cell walls of a plant are removed and the "naked" cells are forced to hybridize, and induced mutagenesis, in which chemicals or irradiation are used to induce random mutations in DNA. The development of these approaches has enhanced the array of techniques that can be used to advance food production. However, as with all other technologies for genetic modification, they also carry the potential for introducing unintended compositional changes that may have adverse effects on human health.

Preventing adverse health effects by maintaining a safe food supply requires the application of appropriate scientific methods to problems of predicting and identifying unintended compositional changes that may result from genetic modification of plants, animals, and microbes intended for consumption as food. To address this need, the U.S. Department of Agriculture, the U.S. Department of Health and Human Services' Food and Drug Administration, and the U.S. Environmental Protection Agency asked the National Academies to convene a committee of scientific experts to outline science-based approaches for assessing or predicting the unintended health effects of genetically engineered (GE) foods and to compare the potential for unintended effects with those of foods derived from other conventional genetic modification methods.

COMMITTEE CHARGE AND APPROACH

This report is intended to aid the sponsoring agencies in evaluating the scientific methods to assess the safety of GE foods before they are sold to the public. The task presented to the committee by the sponsors was to outline science-based approaches to assess or predict unintended health effects of GE foods in order to assist in their evaluation prior to commercialization. The committee was charged to focus on mechanisms by which unintended changes in the biochemical composition of food occur as a result of various conventional and genetic engineering breeding and propagation methods, the extent to which these mechanisms are likely to lead to significant compositional changes in foods that would not be readily apparent without new or enhanced detection methods, and methods to detect such changes in food in order to determine their potential human health effects. The committee was further charged to identify appropriate scientific questions and methods for determining unintended changes in the levels of endogenous nutrients, toxins, toxicants, allergens, or other compounds in food from genetically engineered organisms (GEOs) and outline methods to assess the potential short- and long-term human consequences of such changes.

The committee was charged to compare GE foods with foods derived from other genetic modification methods, such as cross breeding, with respect to the frequency of compositional changes resulting from the modification process and the frequency and severity of the effects of these changes on consumer health. As part of this comparison, the likelihood that elevated toxin or allergen levels would occur in domesticated animals or plants that are modified by different methods was to be considered. Based on this analysis, the committee was charged to discuss whether certain safety issues are specific to GE foods, and if so, recommend approaches for addressing these issues. In addition, the committee was to separately evaluate methods to detect potential unintended compositional changes and health effects of foods derived from cloned animals. The evaluation is presented in a short subreport, separate from, but designed to accompany, the committee's full-length report on foods derived from genetic modification methods.

MECHANISMS BY WHICH UNINTENDED COMPOSITIONAL CHANGES IN FOOD OCCUR AS A RESULT OF BREEDING OR PROPAGATION METHOD

Conventional Breeding

The oldest approach to plant genetic modification is simple selection, where plants exhibiting desired characteristics are selected for continued propagation. Modern technology has improved upon simple selection with the use of molecular analysis to detect plants likely to express desired features. Plants that are selected for desired traits, such as reduced levels of chemicals that produce unpalatable taste, may diminish the ability of plants to survive in the wild because they are also more attractive to pests. Selection for other traits, such as chemicals that increase the resistance of plants to disease, may also be harmful to humans.

Another approach, crossing, can occur within a species or between different species. For example, the generation of triticale, a crop used for both human food and animal feed, arose from the interspecies crossing of wheat and rye. Because most crops can produce allergens, toxins, or antinutritional substances, conventional breeding methods have the potential to produce unintended compositional changes in a food crop.

Genetic Modification

Hazards associated with genetic modifications, specifically genetic engineering, do not fit into a simple dichotomy of genetic engineering versus nongenetic engineering breeding. Not only are many mechanisms common to both genetic engineering as a technique of genetic modification and conventional breeding, but also these techniques slightly overlap each other. Unintentional compositional changes in plants and animals are likely with all conventional and biotechnological breeding methods. The committee assessed the relative likelihood of compositional changes occurring from both genetic engineering and nongenetic engineering modification techniques and generated a continuum to express the potential for unintended compositional changes that reside in the specific products of the modification, regardless of whether the modification was intentional or not (Figure ES-1).

METHODS TO DETECT UNINTENDED CHANGES IN FOOD COMPOSITION

Important advances in analytical methodology for nucleic acids, proteins, and small molecules have occurred over the past decade as a result of concurrent advances in technology and instrumentation; however, there is a need for improvement in all of these areas.

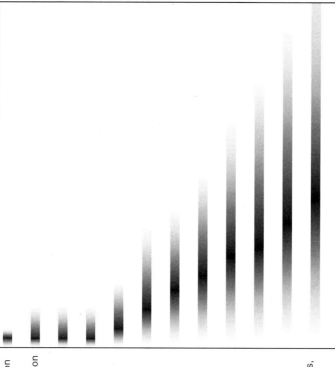

Selection from a homogeneous population

Selection from a heterogeneous population

Crossing of existing approved plant varieties*

Agrobacterium transfer of rDNA from closely related species

Conventional pollen-based crossing of closely related species

Conventional pollen-based crossing of distantly related species and/or embryo rescue

Somatic hybridization

Somaclonal variation (SCV)

Biolistic transfer of rDNA from closely related species

Agrobacterium transfer of rDNA from distantly related species

Biolistic transfer of rDNA from distantly related species

Mutation breeding, chemical mutagenesis, ionizing radiation

Less likely  More likely

*includes all methods of breeding

Currently, there are two basic analytical approaches available to detect compositional changes in food. Targeted quantitative analysis is the traditional approach in which a method is established to quantify a predefined compound or class of compounds. In contrast, profiling methods involve the untargeted analysis of a complex mixture of compounds extracted from a biological sample with the objective of identifying and quantifying all compounds present in a sample. Advanced chemical and genetic profiling techniques—using molecular genetic, proteomic (analysis of complete complements of proteins), and metabolomic (global analysis of nonpeptide small molecules) approaches—are rapidly developing to produce technologies with the potential to provide an enormous amount of data for a given organism, tissue, or food product.

Despite these technological advances in analytical chemistry, our ability to interpret the consequences to human health of changes in food composition is limited. Compositional changes can be readily detected in food and the power of profiling methodologies is rapidly increasing our ability to demonstrate compositional differences among foods. The complexity of food composition challenges the ability of modern analytical chemistry and bioinformatics to chemically identify and determine the biological relevance of the many compositional changes that occur.

METHODS TO ASSESS THE POTENTIAL HUMAN CONSEQUENCES OF UNINTENDED COMPOSITIONAL CHANGES IN FOOD

The major challenges to predicting and assessing unintended adverse health effects of genetically modified (GM) foods—including those that are genetically engineered—are underscored by the severe imbalances between highly advanced analytical technologies and limited abilities to interpret their results and predict health effects that result from the consumption of food that is genetically modified, either by traditional or more modern technologies. The present state of knowledge requires that approaches for assessing the occurrence and significance

FIGURE ES-1   Relative likelihood of unintended genetic effects associated with various methods of plant genetic modification. The gray tails indicate the committee's conclusions about the relative degree of the range of potential unintended changes; the dark bars indicate the relative degree of genetic disruption for each method. It is unlikely that all methods of either genetic engineering, genetic modification, or conventional breeding will have equal probability of resulting in unintended changes. Therefore, it is the final product of a given modification, rather than the modification method or process, that is more likely to result in an unintended adverse effect. For example, of the methods shown, a selection from a homogenous population is least likely to express unintended effects, and the range of those that do appear is quite limited. In contrast, induced mutagenesis is the most genetically disruptive and, consequently, most likely to display unintended effects from the widest potential range of phenotypic effects.

of unintended health effects encompass both targeted and profiling approaches, using a range of toxicological, metabolic, and epidemiological sciences. Encompassing both of these approaches exploits what is known and increases the ability to prevent and assess unsuspected consequences.

Current safety assessments in the premarket period prior to commercialization focus on comparing the GE food with its conventional counterpart to identify uniquely different components. Typically, these comparisons are made on the basis of proximate analysis—an analytical determinant of major classes of food components—as well as nutritional components, toxins, toxicants, antinutrients, and any other characterizing components. The ideal comparator, in most cases, is a near-isogenic variety of food, genetically identical except for the presence of the novel trait, or a near-isogenic parental variety of food from which the GE variety was derived.

In addition to compositional comparisons, agronomic comparisons have been routinely conducted as part of the line selection phase in the development of GE crops. However, these comparisons of phenotypic expression tend to be superficial and could easily miss some varieties containing altered compositions that could impact adversely on human health.

Animal feeding trials are also used to compare the nutritional qualities of a GE crop with its conventional counterpart. Any adverse effects on the health of the animals indicate the possible existence of unexpected alterations in the GE crop that could adversely affect human health, if consumed.

Postmarketing surveillance is an approach to verify premarket screening for unanticipated adverse health consequences from the consumption of GE food. Although postmarketing surveillance has not been used to evaluate any of the GE crops that are currently on the market and there are challenges to its use, this approach holds promise in monitoring potential effects, anticipated and unanticipated, of GE foods that are not substantially equivalent to their conventional counterparts or that contain significantly altered nutritional and compositional profiles.

FRAMEWORK FOR IDENTIFYING AND ASSESSING UNINTENDED ADVERSE EFFECTS FROM GENETICALLY MODIFIED FOODS

The committee developed a framework for a model system based on methods to identify appropriate comparators; increase the knowledge of the determinants of compositional variability; increase the understanding of the biological effects of secondary metabolites in foods; develop more sensitive tools for assessing potential unintended effects from complex mixtures; and improve methods for tracing exposure to GM foods.

The framework, illustrated in a flowchart (Figure ES-2), was used to examine, identify, and evaluate systematically the unintended compositional changes and health effects of GM and, specifically, GE foods. By raising the appropriate questions in this systematic flowchart, the committee has provided a guide for

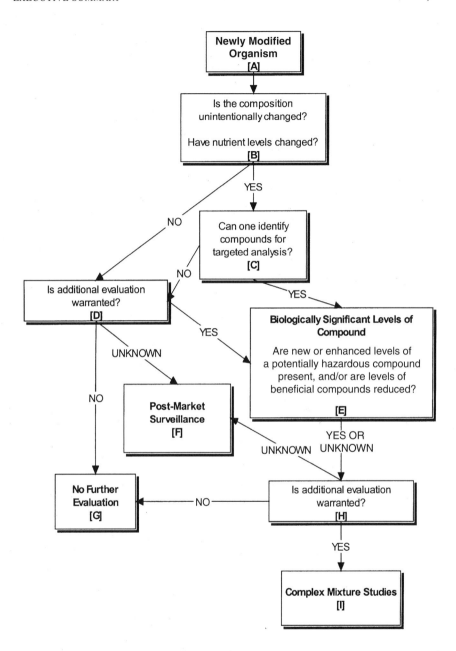

FIGURE ES-2 Flowchart for determining potential unintended effects from genetically modified foods.

overall decision-making, providing alternative routes that can and should be taken according to the specific GM target. Further, the flow chart illustrates the need for appropriate tools to assess and utilize both pre- and postmarket approaches in the process of identifying unintended compositional changes and potential unintended adverse health effects. This model system for selecting and validating methods to detect and assess compositional changes in food serves as the basis for the committee's recommendations to overcome limitations to current methods used to identify compositional differences and evaluate the health significance of new or altered compounds in GM foods.

Overall Findings and Recommendation

Findings

All new crop varieties, animal breeds (see the cloning subreport), and microbial strains carry modified DNA that differs from parental strains. Methods to genetically modify plants, animals, and microbes are mechanistically diverse and include both natural and human-mediated activities. Health outcomes could be associated with the presence or absence of specific substances added or deleted using genetic modification techniques, including genetic engineering, and with unintended compositional changes.

The likelihood that an unintended compositional change will occur can be placed on a continuum that is based on the method of genetic modification used (see Figure ES-1). The genetic modification method used, however, should not be the sole criterion for suspecting and subsequently evaluating possible health effects associated with unintended compositional changes.

All evidence evaluated to date indicates that unexpected and unintended compositional changes arise with all forms of genetic modification, including genetic engineering. Whether such compositional changes result in unintended health effects is dependent upon the nature of the substances altered and the biological consequences of the compounds. To date, no adverse health effects attributed to genetic engineering have been documented in the human population.

Recommendation 1

The committee recommends that compositional changes that result from all genetic modification in food, including genetic engineering, undergo an appropriate safety assessment. The extent of an appropriate safety assessment should be determined prior to commercialization. It should be based on the presence of novel compounds or substantial changes in the levels of naturally occurring substances, such as nutrients that are above or below the normal range for that species (see Chapter 3), taking into account the organism modified and the nature of the introduced trait.

Safety Assessment Tools for Assessing Unintended Effects Prior to Commercialization

Findings

Current voluntary and mandated safety assessment approaches focus primarily on intended and predictable effects of novel components of GE foods. Introduction of novel components into food through genetic engineering can pose unique problems in the selection of suitable comparators for the analytical procedures that are crucial to the identification of unintended compositional changes. Other jurisdictions, particularly the European Union, evaluate all GE food products prior to commercialization, but exempt from similar evaluation all other GM foods. As is discussed in Chapter 3, the policy to assess products based exclusively on their method of breeding is scientifically unjustified.

The most appropriate time for safety assessment of all new food is in the premarket period prior to commercialization, although verification of safety assessments may continue in the postmarket period, generally in cases when a potential problem has been identified or if there is elevated cause for concern. Examples of specific premarket assessments of newly introduced compositional changes to selected GE food are:

- protein, fat, carbohydrate, fiber, ash, and water in a proximate analysis;
- essential macro- and micronutrients in a nutritional analysis;
- known endogenous toxicants and antinutrients in specific species;
- endogenous allergens;
- other naturally occurring, species-specific constituents of potential interest, such as isoflavones and phytoestrogens in soybean or alkaloids in tomato or potato;
- gross agronomic characteristics;
- data derived from domestic animal feeding trials to assess the nutritional quality of new crops; and
- data derived from toxicological studies in animals.

Recommendation 2

The committee recommends that the appropriate federal agencies determine if evaluation of new GM foods for potential adverse health effects from both intended and unintended compositional changes is warranted by elevated concern, such as identification of a novel substance or levels of a naturally occurring substance that exceeds the range of recommended or tolerable intake.

Recommendation 3

For those foods warranting further evaluation, the committee recommends that a safety assessment should be conducted prior to commercialization and continued evaluation postmarket where safety concerns are present. Specifically, the committee recommends the following safety assessment actions.

- Develop a paradigm for identifying appropriate comparators for GE food.
- Collect and make publicly available key compositional information on essential nutrients, known toxicants, antinutrients, and allergens of commonly consumed varieties of food (see the Research Needs section, later in this chapter). These should include mean values and ranges that typically occur as a function of genetic makeup, differences in physiological state, and environmental variables.
- Remove compositional information on GE foods from proprietary domains to improve public accessibility.
- Continue appropriate safety assessments after commercialization to verify premarket evaluations, particularly if the novelty of the introduced substance or the level of a naturally occurring substance leads to increased safety concerns.

Analytical Methodologies

Findings

During the past decade, analytical methodologies for separating and quantifying messenger ribonucleic acids, proteins, and metabolites have improved markedly. Applying these methodologies to the targeted analysis of known nutrients and toxicants will improve the knowledge base for these food constituents. The broad application of targeted methods and continuing development of profiling methods will provide extensive information about food composition and further improve the knowledge base of defined chemical food constituents. The knowledge and understanding needed to relate such compositional information to potential unintended health effects is far from complete, however. Furthermore, currently available bioinformatics and predictive tools are inadequate for correlating compositional analyses with biological effects.

Analytical profiling techniques are appropriate for establishing compositional differences among genotypes, but they must also take into account modification of the profile obtained due to genotype-by-environmental interactions (the influence of the environment on expression of a particular genotype). The knowledge base required to interpret results of profiling methods, however, is insufficiently developed to predict or directly assess potential health effects associated with unintended compositional changes of GM food, as is the necessary associative information (e.g., proteomics, metabolomics, and signaling networks). Additionally, predictive tools to identify the expected behavior of complex and compound structures are limited and require a priori knowledge of their chemical structure, their biological relevance, and their potential interactive targets.

Recommendation 4

The committee recommends the development and employment of standardized sampling methodologies, validation procedures, and performance-based techniques for targeted analyses and profiling of GM food performed in the manner outlined in the flow chart shown in Figure 7-1. Sampling methodology should include suitable comparisons to the near isogenic parental variety of a species, grown under a variety of environmental conditions, as well as ongoing assessment of commonly consumed commercial varieties of food. These include:

• Reevaluation of current methodologies used to detect and assess the biological consequences of unintended changes in GM food, including better tools for toxicity assessment and a more robust knowledge base for determining which novel or increased naturally occurring components of food have a health impact.
• Use of data collection programs, such as the Continuing Survey of Food Intakes by Individuals and the National Health and Nutrition Examination Survey (NHANES), to collect information, prior to commercial release of a new GM food, on current food and nutrient intakes and exposure to known toxins or toxicants through food consumption. The information collected should be used to identify food consumption patterns in the general population and susceptible population subgroups that indicate a potential for adverse reactions to novel substances or increased levels of naturally occurring compounds in GM food.

Additional Tools for Postcommercialization: Identification and Assessment of Unintended Effects

Findings

Postcommercialization or postmarket evaluation tools for verifying and validating premarket assessments of novel substances in food or detectable changes in diet composition, including tracking and epidemiological studies, are important components of the overall assessment of food safety. These tools provide a way to check the efficacy of premarket compositional and safety evaluations through a feedback process. In addition, information databases that result from postmarket studies can be valuable assets in the development of future premarket safety assessment tools.

Postmarket surveillance is a commonly accepted procedure, for example, with new pharmaceuticals and has been beneficial in the identification of harmful and unexpected side effects. As a result, pharmacologists accept postmarket surveillance as a part of the process to identify unexpected adverse outcomes from their products. This example is especially pertinent to GE foods because of the unique ability of this process to introduce gene sequences to generate novel products into organisms intended for use as food and especially in situations where the novel products are introduced at levels that have the potential to alter dietary intake patterns (e.g., elevated levels of key nutrients).

Given the possibility that food with unintended changes may enter the marketplace despite premarket safety mechanisms, postmarket surveillance of exposures and effects is needed to validate premarket evaluations. On the other hand, there are many instances in which postmarket surveillance may not be warranted. For example, when compositional comparisons of a new GM crop or food (e.g., Roundup Ready soybeans) with its conventional counterpart indicate they are compositionally very similar; exposure to novel components remains very low. Thus the process of identifying unintended compositional changes in food is best served by combining premarket testing with postmarket surveillance, when compositional changes indicate that it is warranted, in a feedback loop that follows a new GM food or food product long term, from development through utilization (see Figure ES-2).

Recommendation 5

When warranted by changes such as altered levels of naturally occurring components above those found in the product's unmodified counterpart, population-specific vulnerabilities, or unexplained clusters of adverse health effects, the committee recommends improving the tracking of potential health consequences from commercially available foods that are genetically modified, including those that are genetically engineered, by actions such as the following:

• Improve the ability to identify populations that are susceptible to food allergens and develop databases relevant to tracking the prevalence of food allergies and intolerances in the general population, and in susceptible population subgroups.
• Improve and include other postmarket resources for identifying and tracking unpredicted and unintended health effects from GM foods:
    — Improve the sensitivity of surveys and other analytical methodologies currently used to detect consumer trends in the purchase and use of GM foods after release into the marketplace.
    — Standardize methods for monitoring reports of allergenicity to new foods introduced into the marketplace and apply them to new GM foods.
    — Assure that current food labeling includes relevant nutritional attributes so that consumers can receive more complete information about the nutritional components in GM foods introduced into the marketplace.
    — Improve utilization of potential traceability technology, such as bar coding of animal carcasses and other relevant foods.

• Develop a database of unique genetic sequences (DNA, polymerase chain reaction sequences) from GE foods entering the marketplace to enable their identification in post-market surveillance activities.

• Utilize existing nationwide food intake and health assessment surveys, including NHANES, to:

— Collect comparative information on diet and consumption patterns of the general population and ethnic subgroups in order to account for anthropological differences among population groups and geographic areas where GM foods may be consumed in skewed quantities, recognizing that this will be possible only under selected circumstances where intakes are not evenly distributed across population subgroups of interest and the relevant outcome data are available.

— Provide better representation of the long-term nutritional and other health status information on a full range of children and ethnic groups whose intakes may differ significantly from those of the general population to determine whether changes in health status have occurred as a consequence of consuming novel substances or increased levels of naturally occurring compounds in GM foods released into the marketplace, recognizing again that this will be possible only under selected circumstances that allow one to assess associations between skewed eating patterns and specified health outcomes. Such associations would have to be followed up by other more controlled assessments.

Research Needs

Findings

There is a need, in the committee's judgment, for a broad research and technology development agenda to improve methods for predicting, identifying, and assessing unintended health effects from the genetic modification of food. An additional benefit is that the tools and techniques developed can also be applied to safety assessment and monitoring of foods produced by all methods of genetic modification.

The tools and techniques already developed can be applied to the safety assessment and monitoring of foods produced by all methods of genetic modification. However, although current analytical methods can provide a detailed assessment of food composition, limitations exist in identifying specific differences in composition and interpreting their biological significance.

Recommendation 6

A significant research effort should be made to support analytical methods technology, bioinformatics, and epidemiology and dietary survey tools to detect health changes in the population that could result from genetic modification and, specifically, genetic engineering of food. Specific recommendations to achieve this goal include:

- Focusing research efforts on improving analytical methodology in the study of food composition to improve nutrient content databases and increase understanding of the relationships among chemical components in foods and their relevance to the safety of the food.
- Conducting research to provide new information on chemical identification and metabolic profiles of new GM foods and proteomic profiles on individual compounds and complex mixtures in major food crops and use that information to develop and maintain publicly accessible databases.
- Developing or expanding profiling databases for plants, animals, and microorganisms that are organized by genotype, maturity, growth history, and other relevant environmental variables to improve identification and enhance traceability of GMOs.
- Developing improved bioinformatics tools to aid in the interpretation of food composition data derived from targeting and profiling methods.

Recommendation 7

Research also is needed to determine the relevance to human health of dietary constituents that arise from or are altered by genetic modification. This effort should include:

- Focusing research efforts on developing new tools that can be used to assess potential unintended adverse health effects that result from genetic modification of foods. Such tools should include profiling techniques that relate metabolic components in food with altered gene expression in relevant animal models to specific adverse outcomes identified in GM animal models (animals genetically modified by contemporary biotechnology methods that are proposed to enter the food system).
- Developing improved DNA-based immunological and biochemical tags for selected GM foods entering the marketplace that could be used as surrogate markers to rapidly identify the presence and relative level of specific foods for postmarket surveillance activities.
- Developing improved techniques that enable toxicological evaluations of whole foods and complex mixtures, including:
    — microarray analysis,
    — proteomics, and
    — metabolomics.

CONCLUSION

In response to its charge, the committee has developed a framework to identify appropriate scientific questions and methods for determining unintended changes in the levels of nutrients, toxins, toxicants, allergens, or other compounds in foods from GMOs, in order to assess potential short- and long-term human health consequences of such changes. Although the array of analytical and epidemiological techniques available has increased, there remain sizeable gaps in our ability to identify compositional changes that result from genetic modification of organisms intended for food; to determine the biological relevance of such changes to human health; and to devise appropriate scientific methods to predict and assess unintended adverse effects on human health. The committee has identified and recommended pre- and postmarket approaches to guide assessment of unintended compositional changes that could result from genetic modification of foods and research avenues to fill the knowledge gaps.

The recommendations presented in this report reflect the committee's application of its framework to questions of identification and assessment of unintended adverse health effects from foods produced by all forms of genetic modification, including genetic engineering, and they can serve as a guide for evaluation of future technologies.

1

Introduction

HISTORICAL BACKGROUND

New techniques, collectively referred to as *biotechnology*, have been developed to improve the shelf life, nutritional content, flavor, color, and texture of foods, as well as their agronomic and processing characteristics. One specific biotechnology method is genetic engineering, a type of genetic modification that is the basis for many recent advances in breeding technology (see Appendix A: Glossary, for more comprehensive definitions of key terms used throughout this report). Like any new technology, genetic engineering carries with it some level of uncertainty and requires ways to predict and assess potential unintended effects, whether adverse or beneficial.

Throughout history humans have bred plants, animals, and microbes to achieve traits desirable for different uses. This was done mainly through simple selection and crossbreeding based on the most desired qualities, such as plant vigor, appearance, and taste. It was not until the mid-1800s that scientific understanding of trait inheritance began to emerge. During the 1860s, through his experiments that hybridized different varieties of peas, Gregor Mendel demonstrated the process of heredity (Mendel, 1866). His revolutionary experiments paved the way for modern agriculture by showing that, through controlled pollination crosses, genetic characteristics are inherited in a logical and predictable manner. Since that time many plants have been bred to include desirable traits, such as pest and disease resistance and the ability to overcome environmental stresses. Major gains in crop yields have been attributed partially to advances in these classical plant breeding techniques. Undoubtedly, conventional breeding will continue to play an essential role in improving agricultural crops, domestic animals, and microorganisms used in food production.

17

GENETIC MODIFICATION OF FOOD

Operational Definitions

The terminology used to describe various methods of genetic modification can have different meanings to different readers and can be interpreted in many ways. For the purposes of this report, the committee agreed upon a set of operational definitions for specific terms used to describe methods of genetic modification.

Although in popular parlance the term *genetically modified (GM)* often is used interchangeably with *genetically engineered (GE)* and *biotechnology*, in this report *genetic modification* refers to a range of methods used to alter the genetic composition of a plant or animal, including traditional hybridization and breeding. *Genetic engineering* is one type of genetic modification that involves making an intentional targeted change in a plant or animal gene sequence to effect a specific result (see Figure 1-1) through the use of recombinant deoxyribonucleic acid (rDNA) technology. *Biotechnology* refers to methods (including genetic engineering) other than conventional breeding used to produce new plants, animals, and microbes. *Conventional breeding* is used to describe traditional methods of breeding, or crossing, plants, animals, or microbes with certain desired characteristics for the purpose of generating offspring that express those characteristics.

Overview of Methods to Genetically Modify Plants and Animals

As exemplified by Mendel's research, conventional breeding by crossing has been conducted for centuries to produce genetic modifications in crop plants and farm animals. Even his early experiments, while relatively simple from today's perspective, yielded unexpected results (Mendel, 1866). The concept of dominant inheritance stems from Mendel's unexpected finding that in a cross of white- and red-flowered plants in which the parents were homozygous, the first generation was uniform (F1) but none of the offspring showed an intermediate color, and the second generation (F2) produced three times more red- than white-flowered offspring. This result helped illustrate the distinction between phenotype (physical characteristics) and genotype (genetic pattern).

Plants and animals can be genetically modified in a variety of ways, each requiring some level of human intervention. Traditional methods include selection and crossbreeding, while more contemporary techniques include embryo rescue, cell fusion, somaclonal variation, mutation breeding, and cell selection. Genetic engineering, or rDNA modification, is achieved through different techniques leading to specifically designated genetic changes. There also are methods of genetic manipulation, different from rDNA technology, that use viral vectors to introduce foreign DNA into host cells.

All methods of genetic modification hold the potential, either intentionally or unintentionally, to alter levels of primary metabolites (such as proteins, lipids,

FIGURE 1-1   A general history of genetics and genetic modification.
Source: Adapted from The Arabidopsis Initiative (2000); Moore (2003); University of Illinois (1999).

and carbohydrates) and a wide variety of secondary metabolites with either beneficial or adverse results. For example, introducing new proteins or increasing the levels of endogenous proteins in a food product may increase its potential for allergenicity, but genetic engineering may also reduce the allergenicity of a plant used for food or reduce its levels of known toxins.

Given the diverse assortment of techniques used to genetically modify plants and animals, it is clear that unintended adverse health effects potentially associated with these techniques do not fit a simple dichotomy of comparing genetic engineering with traditional breeding.

An important step for determining the likelihood of such unintended adverse health effects is assessing the compositional similarity between a conventional plant or animal and its genetically modified counterpart. Attributes of genetically engineered organisms (GEOs) typically compared with those of their traditionally bred counterparts include gene sources, phenotypic characteristics (such as size, shape, and color), composition (such as nutrients, antinutrients, allergens), and consumption patterns. Additional safety studies may be conducted, focusing on areas of greatest potential concern.

Comparing a GE plant or animal with its conventional counterpart alone is not sufficient for assessing the likelihood of unintended effects of genetic engineering and conventional breeding practices. It also is necessary to determine the frequency and nature of the associated unintended effects and to evaluate the methods that are potentially useful in assessing the safety of food products that result from use of these methods.

The Scope of This Report

While using biotechnology or conventional breeding techniques to enhance specific characteristics or increase the yield of food introduces the possibility of unintended deleterious effects on both human health and the environment, the focus of this report is health—including an examination of whether the likelihood of unintended adverse health effects from compositional changes is greater for foods that are genetically engineered than for those genetically modified using other methods (such as conventionally bred plants). Furthermore, this report evaluates currently used and newly developed methods for detecting unintended changes in genetically modified foods and also assesses and recommends techniques for predicting their potential health effects. However, it does not directly evaluate the potential health effects of specific engineered genes or proteins, nor does it assess the regulation of GE food.

THE CHARGE TO THE COMMITTEE

Three federal government agencies—the U.S. Department of Agriculture, the U.S. Department of Health and Human Services' Food and Drug Administration, and the U.S. Environmental Protection Agency—asked the National Academies to convene a committee that would outline science-based approaches to assess or predict the unintended health effects of GE foods to aid in evaluating these products before they are sold to the public. The committee was charged with identifying appropriate scientific questions and methods for determining unintended changes in the levels of nutrients, toxins, toxicants, allergens, or other compounds in food from GEOs and outlining methods to assess the potential short- and long-term human health consequences of such changes.

The agencies also asked the committee to compare GE food with food derived from other genetic modification methods, such as crossbreeding, with respect to the frequency of compositional changes and the frequency and severity of the effects of these changes on consumer health. Finally, the committee was asked to discuss whether certain safety issues are specific to GE food and, if so, to recommend approaches for addressing these issues.

The committee's charge did not include evaluating or making recommendations about policy issues, such as labeling GE foods, segregating foods in commerce, or preventing cross-contamination of foods.

Approach to the Task

The committee approached its task by gathering information from existing literature and from public workshop presentations by recognized experts (see Appendix B for the workshop agendas) and then deliberating on issues relevant to their charge.

From these discussions, the committee developed a theoretical framework for identifying appropriate comparators for GE and other GM foods, increasing scientific understanding of the determinants of compositional variability among foods, increasing understanding of the biological effects of secondary metabolites in food, developing more sensitive techniques for assessing potential unintended effects from food modification, and improving methods for tracking and tracing exposure in genetically modified food.

The committee's deliberations about identifying appropriate comparators for GE food clarified that while such comparisons are necessary, they alone are not sufficient for determining the likelihood of producing an unintended adverse health effect. Consequently, this report focuses on an array of complementary science-based approaches for predicting and assessing unintended health effects of GE food and for evaluating the mechanisms by which unintended effects occur as a result of genetic modification.

Organization of the Report

This report is organized into seven chapters and an accompanying subreport on animal genetic manipulation and cloning. Chapter 2 describes the molecular biological and biochemical methods of genetic manipulation of plants, animals, and microorganisms. Chapter 3 discusses the potential for unintended compositional changes from different methods of breeding. Chapter 4 outlines new approaches for identifying unintended changes in food composition. Chapter 5 details diverse ways that adverse health effects can occur from food, while Chapter 6 suggests methods for predicting and assessing those effects that result from intended and unintended compositional changes resulting from genetic modification. Chapter 7 presents the committee's conclusions and recommendations. The subreport on animal genetic manipulation and cloning reviews the current literature and makes recommendations for methodologies that could be used to assess cloned animal products.

REFERENCES

The Arabidopsis Initiative. 2000. Analysis of the genome sequence of the flowering plant *Arabidopsis thaliana*. *Nature* 408:796–815.

Mendel G. 1866. Experiments in plant-hybridization. *Verh Naturforsch Ver Brunn Abh* 4. Translated by Bateson W, reprinted in Peters J. 1959. *Classic Papers in Genetics*. Englewood Cliffs, NJ: Prentice-Hall. Pp. 2–20.

Moore G. 2003. *Timeline of Plant Tissue Culture and Selected Molecular Biology Events*. Online. University of Florida Institute for Food and Agricultural Sciences. Available at http://www.hos.ufl.edu/mooreweb/TissueCulture/class2/Timeline%20of%20Plant%20Tissue%20Culture%20and%20Selected%20Molecular%20Biology%20Events.doc. Accessed August 29, 2003.

University of Illinois. 1999. *The Economics and Politics of GMOs in Agriculture*. Online. College of Agricultural, Consumer, and Environmental Science, Bulletin 809. Available at http://web.aces.uiuc.edu/wf/GMOs.htm. Accessed August 29, 2003.

2

Methods and Mechanisms for Genetic Manipulation of Plants, Animals, and Microorganisms

This chapter provides a brief description of genetic modification methods used to develop new plant, animal, and microbial strains for use as human food. The next chapter (Chapter 3) presents a detailed analysis of the likelihood for these methods to result in unintentional compositional changes.

BACKGROUND

Modification to produce desired traits in plants, animals, and microbes used for food began about 10,000 years ago. These changes, along with natural evolutionary changes, have resulted in common food species that are now genetically different from their ancestors.

Advantageous outcomes of these genetic modifications include increased food production, reliability, and yields; enhanced taste and nutritional value; and decreased losses due to various biotic and abiotic stresses, such as fungal and bacterial pathogens. These objectives continue to motivate modern breeders and food scientists, who have designed newer genetic modification methods for identifying, selecting, and analyzing individual organisms that possess genetically enhanced features.

For plant species, it can take up to 12 years to develop, evaluate, and release a new variety of crop in accordance with international requirements, which specify that any new variety must meet at least three criteria: it must be genetically distinct from all other varieties, it must be genetically uniform through the population, and it must be genetically stable (UPOV, 2002).

While advances in modification methods hold the potential for reducing the time it takes to bring new foods to the marketplace, an important benefit of a long

evaluation period is that it provides opportunities for greater assurance that deleterious features will be identified and potentially harmful new varieties can be
eliminated before commercial release. As discussed more fully in Chapter 5, it is
both prudent and preferable to identify potentially hazardous products before they
are made commercially available, and with few exceptions standard plant breeding practices have been very successful in doing so.

PLANT GENETIC MODIFICATION

Techniques Other than Genetic Engineering

Simple Selection

The easiest method of plant genetic modification (see Operational Definitions in Chapter 1), used by our nomadic ancestors and continuing today, is simple
selection. That is, a genetically heterogeneous population of plants is inspected,
and "superior" individuals—plants with the most desired traits, such as improved
palatability and yield—are selected for continued propagation. The others are
eaten or discarded. The seeds from the superior plants are sown to produce a new
generation of plants, all or most of which will carry and express the desired traits.
Over a period of several years, these plants or their seeds are saved and replanted,
which increases the population of superior plants and shifts the genetic population so that it is dominated by the superior genotype. This very old method of
breeding has been enhanced with modern technology.

An example of modern methods of simple selection is *marker-assisted selection*, which uses molecular analysis to detect plants likely to express desired features, such as disease resistance to one or more specific pathogens in a population. Successfully applying marker-assisted selection allows a faster, more
efficient mechanism for identifying candidate individuals that may have "superior traits."

Superior traits are those considered beneficial to humans, as well as to domesticated animals that consume a plant-based diet; they are not necessarily beneficial to the plant in an ecological or evolutionary context. Often traits considered beneficial to breeders are detrimental to the plant from the standpoint of
environmental fitness. For example, the reduction of unpalatable chemicals in a
plant makes it more appealing to human consumers but may also attract more
feeding by insects and other pests, making it less likely to survive in an
unmanaged environment. As a result, cultivated crop varieties rarely establish
populations in the wild when they escape from the farm. Conversely, some traits
that enhance a plant's resistance to disease may also be harmful to humans.

Crossing

Crossing occurs when a plant breeder takes pollen from one plant and brushes it onto the pistil of a sexually compatible plant, producing a hybrid that carries genes from both parents. When the hybrid progeny reaches flowering maturity, it also may be used as a parent.

Plant breeders usually want to combine the useful features of two plants. For example, they might add a disease-resistance gene from one plant to another that is high-yielding but disease-susceptible, while leaving behind any undesirable genetic traits of the disease-resistant plant, such as poor fertility and seed yield, susceptibility to insects or other diseases, or the production of antinutritional metabolites.

Because of the random nature of recombining genes and traits in crossed plants, breeders usually have to make hundreds or thousands of hybrid progeny to create and identify those few that possess useful features with a minimum of undesirable features. For example, the majority of progeny may show the desired disease resistance, but unwanted genetic features of the disease-resistant parent may also be present in some. Crossing is still the mainstay of modern plant breeding, but many other techniques have been added to the breeders' tool kit.

Interspecies Crossing

Interspecies crossing can take place through various means. Closely related species, such as cultivated oat (*Avena sativa*) and its weedy relative wild oat (*Avena fatua*), may cross-pollinate for exchange of genetic information, although this is not generally the case. Genes from one species also can naturally integrate into the genomes of more distant relatives under certain conditions. Some food plants can carry genes that originate in different species, transferred both by nature and by human intervention. For example, common wheat varieties carry genes from rye. A common potato, *Solanum tuberosum*, can cross with relatives of other species, such as *S. acaule* (Kozukue et al., 1999) or *S. chacoense* (Sanford et al., 1998; Zimnoch-Guzowska et al., 2000).

Chromosome engineering is the term given to nonrecombinant deoxyribonucleic acid (rDNA) cytogenetic manipulations, in which portions of chromosomes from near or distant species are recombined through a natural process called chromosomal translocation. Sears (1956, 1981) pioneered the human exploitation of this process, which proved valuable for transferring traits that were otherwise unattainable, such as pest or disease resistance, into crop species. However, because transferring large segments of chromosomes also transferred a number of neutral or detrimental genes, the utility of this technique was limited.

Recent refinements allow plant breeders to restrict the transferred genetic material, focusing more on the gene of interest (Lukaszewski, 2004). As a result, chromosome engineering is becoming more competitive with rDNA technology in its ability to transfer relatively small pieces of DNA. Several crop species, such as corn, soybean, rice, barley, and potato, have been improved using chromosome engineering (Gupta and Tsuchiya, 1991).

Embryo Rescue

Sometimes human technical intervention is required to complete an interspecies gene transfer. Some plants will cross-pollinate and the resulting fertilized hybrid embryo develops but is unable to mature and sprout. Modern plant breeders work around this problem by pollinating naturally and then removing the plant embryo before it stops growing, placing it in a tissue-culture environment where it can complete its development. Such embryo rescue is not considered genetic engineering, and it is not commonly used to derive new varieties directly, but it is used instead as an intermediary step in transferring genes from distant, sexually incompatible relatives through intermediate, partially compatible relatives of both the donor and recipient species.

Somatic Hybridization

Recent advances in tissue-culture technologies have provided new opportunities for recombining genes from different plant sources. In *somatic hybridization,* a process also known as *cell fusion,* cells growing in a culture medium are stripped of their protective walls, usually using pectinase, cellulase, and hemicellulase enzymes. These stripped cells, called *protoplasts*, are pooled from different sources and, through the use of varied techniques such as electrical shock, are fused with one another.

When two protoplasts fuse, the resulting somatic hybrid contains the genetic material from both plant sources. This method overcomes physical barriers to pollen-based hybridization, but not basic chromosomal incompatibilities. If the somatic hybrid is compatible and healthy, it may grow a new cell wall, begin mitotic divisions, and ultimately grow into a hybrid plant that carries genetic features of both parents. While protoplast fusions are easily accomplished, as almost all plants (and animals) have cells suitable for this process, relatively few are capable of regenerating a whole organism, and fewer still are capable of sexual reproduction. This non-genetic engineering technique is not common in plant breeding as the resulting range of successful, fertile hybrids has not extended much beyond what is possible using other conventional technologies.

Somaclonal Variation

Somaclonal variation is the name given to spontaneous mutations that occur when plant cells are grown in vitro. For many years plants regenerated from tissue culture sometimes had novel features. It was not until the 1980s that two Australian scientists thought this phenomenon might provide a new source of genetic variability, and that some of the variant plants might carry attributes of value to plant breeders (Larkin and Scowcroft, 1981).

Through the 1980s plant breeders around the world grew plants in vitro and scored regenerants for potentially valuable variants in a range of different crops. New varieties of several crops, such as flax, were developed and commercially released (Rowland et al., 2002). Molecular analyses of these new varieties were not required by regulators at that time, nor were they conducted by developers to ascertain the nature of the underlying genetic changes driving the variant features. Somaclonal variation is still used by some breeders, particularly in developing countries, but this non-genetic engineering technique has largely been supplanted by more predictable genetic engineering technologies.

Mutation Breeding: Induced Chemical and X-ray Mutagenesis

Mutation breeding involves exposing plants or seeds to mutagenic agents (e.g., ionizing radiation) or chemical mutagens (e.g., ethyl methanesulfonate) to induce random changes in the DNA sequence. The breeder can adjust the dose of the mutagen so that it is enough to result in some mutations, but not enough to be lethal. Typically a large number of plants or seeds are mutagenized, grown to reproductive maturity, and progeny are derived. The progeny are assessed for phenotypic expression of potentially valuable new traits.

As with somaclonal variation, the vast majority of mutations resulting from this technique are deleterious, and only chance determines if any genetic changes useful to humans will appear. Other than through varying the dosage, there is no means to control the effects of the mutagen or to target particular genes or traits. The mutagenic effects appear to be random throughout the genome and, even if a useful mutation occurs in a particular plant, deleterious mutations also will likely occur. Once a useful mutation is identified, breeders work to reduce the deleterious mutations or other undesirable features of the mutated plant. Nevertheless, crops derived from mutation breeding still are likely to carry DNA alterations beyond the specific mutation that provided the superior trait.

Induced-mutation crops in most countries (including the United States) are not regulated for food or environmental safety, and breeders generally do not conduct molecular genetic analyses on such crops to characterize the mutations or determine their extent. Consequently, it is almost certain that mutations other than those resulting in identified useful traits also occur and may not be obvious, remaining uncharacterized with unknown effects.

Worldwide, more than 2,300 different crop varieties have been developed using induced mutagenesis (FAO/IAEA, 2001), and about half of these have been developed during the past 15 years. In the United States, crop varieties ranging from wheat to grapefruit have been mutated since the technique was first used in the 1920s. There are no records of the molecular characterizations of these mutant crops and, in most cases, no records to retrace their subsequent use.

Cell Selection

Several commercial crop varieties have been developed using *cell selection,* including varieties of soybeans (Sebastian and Chaleff, 1987), canola (Swanson et al., 1988), and flax (Rowland et al., 1989). This process involves isolating a population of cells from a so-called "elite plant" with superior agricultural characteristics. The cells are then excised and grown in culture. Initially the population is genetically homogeneous, but changes can occur spontaneously (as in somaclonal variation) or be induced using mutagenic agents. Cells with a desired phenotypic variation may be selected and regenerated into a whole plant. For example, adding a suitable amount of the appropriate herbicide to the culture medium may identify cells expressing a novel variant phenotype of herbicide resistance. In theory, all of the normal, susceptible cells will succumb to the herbicide, but a newly resistant cell will survive and perhaps even continue to grow. An herbicide-resistant cell and its derived progeny cell line thus can be selected and regenerated into a whole plant, which is then tested to ensure that the phenotypic trait is stable and results from a heritable genetic alteration. In practice, many factors influence the success of the selection procedure, and the desired trait must have a biochemical basis that lends itself to selection in vitro and at a cellular level.

Breeders cannot select for increased yield in cell cultures because the cellular mechanism for this trait is not known. The advantage of cell selection over conventional breeding is the ability to inexpensively screen large numbers of cells in a petri dish in a short time instead of breeding a similar number of plants in an expensive, large field trial conducted over an entire growing season.

Like somaclonal variation, cell selection has largely been superceded by recombinant technologies because of their greater precision, higher rates of success, and fewer undocumented mutations.

Genetic Engineering

As noted in Chapter 1, this report defines *genetic engineering* specifically as one type of genetic modification that involves an intended targeted change in a plant or animal gene sequence to effect a specific result through the use of rDNA technology. A variety of genetic engineering techniques are described in the following text.

Microbial Vectors

Agrobacterium tumefaciens is a naturally occurring soil microbe best known for causing crown gall disease on susceptible plant species. It is an unusual pathogen because when it infects a host, it transfers a portion of its own DNA into the plant cell. The transferred DNA is stably integrated into the plant DNA, and the plant then reads and expresses the transferred genes as if they were its own. The transferred genes direct the production of several substances that mediate the development of a crown gall.

Among these substances is one or more unusual nonprotein amino acids, called opines. Opines are translocated throughout the plant, so food developed from crown gall-infected plants will carry these opines. In the early 1980s strains of *Agrobacterium* were developed that lacked the disease-causing genes but maintained the ability to attach to susceptible plant cells and transfer DNA.

By substituting the DNA of interest for the crown gall disease-causing DNA, scientists derived new strains of *Agrobacterium* that deliver and stably integrate specific new genetic material into the cells of target plant species. If the transformed cell then is regenerated into a whole fertile plant, all cells in the progeny also carry and may express the inserted genes. *Agrobacterium* is a naturally occurring genetic engineering agent and is responsible for the majority of GE plants in commercial production.

Microprojectile Bombardment

Klein and colleagues (1987) discovered that naked DNA could be delivered to plant cells by "shooting" them with microscopic pellets to which DNA had been adhered. This is a crude but effective physical method of DNA delivery, especially in species such as corn, rice, and other cereal grains, which *Agrobacterium* does not naturally transform. Many GE plants in commercial production were initially transformed using microprojectile delivery.

Electroporation

In *electroporation,* plant protoplasts take up macromolecules from their surrounding fluid, facilitated by an electrical impulse. Cells growing in a culture medium are stripped of their protective walls, resulting in protoplasts. Supplying known DNA to the protoplast culture medium and then applying the electrical pulse temporarily destabilizes the cell membrane, allowing the DNA to enter the cell. Transformed cells can then regenerate their cell walls and grow to whole, fertile transgenic plants. Electroporation is limited by the poor efficiency of most plant species to regenerate from protoplasts.

Microinjection

DNA can be injected directly into anchored cells. Some proportion of these cells will survive and integrate the injected DNA. However, the process is labor intensive and inefficient compared with other methods.

Transposons/Transposable Elements

The genes of most plant and some animal (e.g., insects and fish) species carry transposons, which are short, naturally occurring pieces of DNA with the ability to move from one location to another in the genome. Barbara McClintock first described such transposable elements in corn plants during the 1950s (Cold Spring Harbor Laboratory, 1951). Transposons have been investigated extensively in research laboratories, especially to study mutagenesis and the mechanics of DNA recombination. However, they have not yet been harnessed to deliver novel genetic information to improve commercial crops.

Nontransgenic Molecular Methods of Manipulation

Genetic features can be added to plants and animals without inserting them into the recipient organism's native genome. DNA of interest may be delivered to a plant cell, expressing a new protein—and thereby a new trait—without becoming integrated into the host-cell DNA. For example, virus strains may be modified to carry genetic material into a plant cell, replicate, and thrive without integrating into the host genome. Without integration, however, new genetic material may be lost during meiosis, so that seed progeny may not carry or express the new trait.

Many food plants are perennials or are propagated by vegetative means, such as grafting or from cuttings. In these cases the virus and new genes would be maintained in subsequent, nonsexually generated populations. Technically such plants are not products of rDNA because there is no recombination or insertion of introduced DNA into the host genome. Although these plants are not GE, they do carry new DNA and new traits. No such products are known to be currently on the market in the United States or elsewhere. (See McHughen [2000] for further information on genetic mechanisms used in plant improvement.)

ANIMAL GENETIC MODIFICATION

Techniques Other than Genetic Engineering

Domestication and Artificial Selection

Modern breeds of livestock differ markedly from their ancestors as a result of breeding strategies. For example, milk production per cow has increased among Holstein dairy cattle. Similarly, breeding programs have resulted in lean, fast-growing pigs (Notter, 1999). Chickens from modern breeds each produce more than 250 eggs per year, approximately double that produced in 1950, again mainly due to genetic selection.

Established and emerging biotechnologies in animal agriculture include assisted reproductive technologies; use of naturally occurring hormones, such as recombinant bovine somatotropin; marker-assisted selection; biotechnologies to enhance reproductive efficiency without affecting the genome; and biotechnologies to enhance expression of desirable genes.

Assisted Reproductive Procedures

Modern breeds of livestock differ from their ancestors because the use of frozen semen for artificial insemination (AI), along with sire testing and sire selection, has markedly affected the genetic quality of livestock, especially dairy cattle. Select bulls are tested for fertility and judged on the basis of the milk that their daughters produce. A notable example is the milk from Holstein cows, which increased almost threefold between 1945 and 1995 (Majeskie, 1996) through a combination of AI using semen from select bulls and improved milk production management (Diamond, 1999; Hale, 1969). Using sophisticated statistical models to predict breeding values, sire testing and selection, crossbreeding, and marker-assisted selection, along with AI, have greatly advanced the production characteristics of livestock. It is expected that AI will continue to be an integral tool in animal production systems.

(Assisted reproductive and recombinant hormone technologies are discussed in detail in the accompanying subreport, *Methods and Mechanisms of Genetic Manipulation and Cloning of Animals.*)

Techniques Fundamental to Genetic Engineering in Livestock

Although the following are not methods to generate modifications per se, they are considered modern methods that support the overall breeding and selection system for propagating desired genotypes for animals expressing desired traits.

Embryo Recovery and Transfer and Superovulation

Embryo recovery and transfer allow valuable animals to contribute more offspring to the gene pool (Seidel, 1984). Embryos that are frozen and stored before being used to initiate a pregnancy result in 40,000 to 50,000 beef calves per year (NAAB, 2000). Emerging technologies will allow the sexing of semen and embryos to control the gender of the offspring. The production of single-sex sperm, by cell sorting X and Y sperm, will greatly benefit the livestock industries (Johnson, 2000).

In Vitro Maturation and Fertilization of Oocytes

Up to several thousand embryos can be produced using techniques for recovering and maturing immature eggs, or oocytes, in about one day in a medium containing hormones, and then fertilizing them with live sperm or injecting a single sperm or sperm head into their outer layers—either beneath the zona pellucida or directly into the cytoplasm. The resulting zygotes are cultured in vitro, usually to the blastocyst stage, before being transferred to recipient females (First, 1991). The commercial application of in vitro maturation and fertilization has resulted in as many as 4,000 calves being born in a single year (NAAB, 2000).

Embryo Splitting

Splitting or bisecting embryos yields zygotic twins, or non-GE clones, that are genetically identical in both their nuclear and mitochondrial genes (Heyman et al., 1998). Maternal twins exhibit greater variation in phenotype than paternal twins with only one X chromosome. Further, there is the potential for differences in mitochondrial DNA distribution to affect phenotype.

These embryos are then placed in an empty zona pellucida and transferred to recipient females, which carry them to term. Through 2001, a total of 2,226 registered Holstein clones—754 males and 1,472 females—were produced from embryo splitting, with 1 to 2 percent of calves produced (NAAB, 2000).

Genetic Engineering

Cloning as a technique, and the implications for predicting and assessing adverse health effects that may be associated with this technique, are addressed in the committee's subreport that follows this report.

Techniques employed to introduce novel genes into domestic animals are discussed in detail in the report *Animal Biotechnology: Science Based Concerns* (NRC, 2002). These transgenic approaches applicable to animals are summarized in the following text.

Accessing the Germline of Animals

Germline refers to the lineage of cells that can be genetically traced from parent to offspring. It is possible to access the germline of animals using one of five methods (NRC, 2002):

1. directly manipulating the fertilized egg after it has been implanted in the uterus;
2. manipulating the sperm that produces the zygote;
3. manipulating early embryonic tissue in place;
4. using embryonic stem cell lines in early embryos; and
5. manipulating cultured somatic cells to transfer their nuclei into enucleated oocytes.

Transfection

Several of the methods used to transfect or introduce novel genes into animals are similar to those used for plants. Commonly used methods include:

- microinjection of DNA into the nucleus of anchored cells;
- electroporation, where DNA is introduced through cell membrane pores by pulsed electrical charges;
- polycationic neutralization of the cell membrane and the DNA to be introduced to improve passive uptake;
- lipofection, where DNA is; and
- sperm-mediated transfection, often used in conjunction with intracytoplasmic sperm injection or electroporation.

As is the case with plants, microinjection is a highly inefficient means of creating transgenic animals. For example, an incredibly small percentage of livestock embryos that undergo microinjection yield transgenic animals (Rexroad, 1994). Moreover, successfully microinjected transgenic animals do not necessarily pass their transgenes on to their offspring (NRC, 2002).

Retroviral Vectors

This method is similar to viral delivery methods used in plants in that virus strains are modified to carry genetic material into a cell. It differs in that after the novel DNA is delivered, the viral replication process integrates it into the host cell's genome.

Transposons

The use of transposable elements in animal cells has not been completely developed. Although no active naturally occurring transposable elements have been found in mammals, those found in insects and fish are under investigation for potential use in animals.

Knock-In and Knock-Out Technology

Transgenic technology can also be used to create organisms that lack specific genes or those in which one existing gene has been replaced by another that has been engineered. The addition ("knock-in") or deletion ("knock-out") of specific gene functions through introduced mutations or genetic engineering based on homologous recombination has become commonplace in animals used for experimentation, such as mice. Although at present this technology is not efficient and thus not practical for use in generating knock-in or knock-out domestic animals, there are examples of its use in domestic sheep and pigs. (NRC, 2002).

Marker-Assisted Selection

Marker-assisted selection involves establishing a link between inheriting a desirable trait, such as milk yield, and segregating specific genetic markers that are coupled to that trait. Marker-assisted selection is important in animal breeding and selection strategies for studying complex traits governed by many genes (Georges, 2001). The use of this method is expected to increase exponentially as genome-sequencing projects identify greater numbers of useful, segregated markers for economically important traits.

Initially animals will be screened for genes that control simple traits that may be undesirable, such as horns in cattle or metabolic stress syndrome in pigs. In time, easily identifiable markers that accompany multiple genes controlling more complex traits, such as meat tenderness and taste, growth, offspring size, and disease resistance, will become available to improve animal health and production traits (Dekkers and Hospital, 2002).

Two notable examples can be found in sheep. One is the Booroola gene in which a single-nucleotide base change is responsible for the callipyge muscle hypertrophy phenotype—the only known example of polar over-dominance in a mammal (Freking et al., 2002). Another is introgression of the Booroola gene into the Awassi and the Assaf dairy breeds (Gootwine, 2001).

Sequencing genomes of animals that are important to agriculture will identify genes that influence reproductive efficiency. For example, a growth-hormone-receptor variant on bovine chromosome 20 affects the yield and composition of milk, and is expected to increase milk production by 200 kg per lactation and decrease milk fat from 4.4 percent to 3.4 percent (Fletcher, 2003).

Nontransgenic Methods of Animal Manipulation

Biotechnology can be used to modify endocrine function of domestic animals and affect reproduction, lactation, and growth. For example, in pigs and rats (Draghia-Akli et al., 2002) hypothalamic-specific expression of growth-hormone-releasing hormone is not essential since ectopic expression of a cloned DNA for this neuropeptide can be genetically driven by a synthetic muscle-specific transcriptional promoter to elicit increases in both growth hormone and insulin-like growth factor-I (Khan et al., 2002). This biotechnology has the potential, by using specific hormones and growth factors during critical developmental periods, to enhance uterine capacity and to increase milk production.

GENETIC MODIFICATION OF MICROBES

Humans have used and genetically modified (GM) microbes for centuries to produce food. Wine, bread, and cheese are common examples of ancient foods, still popular today, that depend on microbial ingredients and activities. Endogenous populations of microbes, particularly bacteria and yeasts, are genetically varied enough to provide sufficiently different traits to allow the development of useful microbial strains through simple selection or induced mutation.

Microorganisms play significant roles in food production. They serve primary and secondary roles in food fermentation and in food spoilage, and they can produce enzymes or other metabolites used in food production and processing. Fermentations can be initiated and conducted completely by the bacterial populations that are endogenous to the raw materials being fermented. However, it is more reliable in terms of uniformity and predictability to intentionally introduce starter cultures to initiate the fermentation and, in some instances, to perform the complete fermentation process. Most fermented products now are prepared this way in industrialized countries.

The types of microorganisms that carry out food fermentations range from bacteria to molds and yeasts, but by far the most widely used organisms are lactic acid bacteria (LAB) and yeasts (*Sacchromyces cerevisiae*). Traditional genetic modification methods that have been employed—particularly for microbial starter cultures—include selection, mutagenesis, conjugation, and protoplast fusion, the last of which is analogous to somatic hybridization in plant systems.

Before molecular genetics was developed and applied to LAB, the most widely used genetic modification method was chemical- or ultra-violet-induced mutagenesis, followed by an enrichment or selection process for mutants with superior characteristics.

A second traditional approach, conjugation, relies on natural methods of genetic exchange whereby DNA is transferred from one strain to another. Conjugation can occur between LAB strains as well as between LAB and other bacteria (Steenson and Klaenhammer, 1987). Although the resulting strains could conceivably be labeled as recombinant, the fact that this process can occur naturally circumvents application of the GE organism's classification.

A less common, but still useful, method has been to use protoplast fusion to facilitate recombination between two strains with superior but unique characteristics, producing a strain that possesses the desired characteristics of both parents. Protoplast fusion was classically used as a mapping method in bacteria and only recently has been used successfully to produce strains of LAB with desired characteristics (Patnaik et al., 2002). It has, however, been successfully used for some time to generate yeast strains that produce a greater number of biochemical substrates for use in the fermentation process (Pina et al., 1986).

Given the number and combinations of desirable traits in starter culture organisms, producers have remained interested in developing improved starter cultures, using essentially two different approaches. The traditional approach has been to identify endogenous strains with desirable traits by conducting many small-scale fermentations. This type of trial-and-error approach is far from practical because, while productive, low throughput is a limiting factor in the success rate.

The second approach is to produce the desired traits in the laboratory using molecular genetic and genetic engineering techniques. With the burgeoning field of genomics and the public availability of hundreds of fully sequenced bacterial genomes, this approach has become highly attractive and efficient and is favored by industry. Its primary advantage is the precision with which starter culture strains can be engineered.

The most common method used to introduce recombinant DNA into microorganisms is *transformation,* whereby DNA of interest is introduced directly into recipient cells by making them permeable using chemical agents, enzymes, or electroporation. The first method developed for LAB was plasmid protoplast fusion, in which recipient cells are stripped of walls and subsequently fused with polyethylene glycol, trapping the newly introduced DNA between the cells (Kondo and McKay, 1984).

Electroporation was developed for LAB during the late 1980s and employs electrical currents to create pores in the cell envelope, allowing DNA from other sources to enter (Luchansky et al., 1988). This method is probably the most widely used for research due to its simplicity. However, it lacks efficiency in many different species.

Recombinant DNA also can be introduced into LAB using a technique called *transduction,* in which a bacteriophage is used to move DNA from one strain into another (Bierkland and Holo, 1993). Unlike transformation, transduction can be fraught with problems that cause deletions within the plasmid (known as transductional shortening that are typically of undefined length).

Microbial transformation is usually simpler and more efficient than transformation in higher organisms, and has been in use longer for the development of commercial strains. Academic research also has been able to scrutinize the molecular genetic effects of transformation in microbes to a much greater extent than it has in higher organisms. Principles gleaned from studies of microbes have proven instrumental in understanding analogous events in the molecular genetics of higher organisms.

REFERENCES

Bierkland NK, Holo H. 1993. Transduction of a plasmid carrying the cohesive end region from *Lactococcus lactis* bacteriophage LC3. *Appl Environ Microbiol* 59:1966–1968.

Cold Spring Harbor Laboratory. 1951. *Cold Spring Harbor Symposium on Quantitative Biology XVI: Genes and Mutations.* Online. Available at http://library.cshl.edu/symposia/1951/index.html. Accessed November 4, 2003.

Dekkers JC, Hospital F. 2002. The use of molecular genetics in the improvement of agricultural populations. *Nat Rev Genet* 3:22–32.

Diamond J. 1999. *Guns, Germs, and Steel: The Fates of Human Societies.* New York: WW Norton.

Draghia-Akli R, Malone PB, Hill LA, Ellis KM, Schwartz RJ, Nordstrom JL. 2002. Enhanced animal growth via ligand-regulated GHRH mygenic-injectable vectors. *FASEB J* 16:426–428.

FAO/IAEA (Food and Agriculture Organization of the United Nations/International Atomic Energy Agency). 2001. *FAO/IAEA Mutant Varieties Database.* Online. Available at http://www-infocris.iaea.org/MVD/. Accessed January 1, 2003.

First NL. 1991. New advances in reproductive biology of gametes and embryos. In: Petersen RA, McLaren A, First NL, eds. *Animal Applications of Research in Mammalian Development: Current Communications in Cell and Molecular Biology.* Cold Spring Harbor, NY: Cold Spring Harbor Laboratory Press. Pp. 1–21.

Fletcher A, ed. 2003. *Gene Identified to Regulate Milk Content and Yield.* Online. FoodProductionDaily.com. Available at http://foodproductiondaily.com/news/news-NG.asp?id=29318. Accessed February 28, 2003.

Freking BA, Murphy SK, Wylie AA, Rhodes SJ, Keele JW, Leymaster KA, Jirtle RL, Smith TP. 2002. Identification of the single base change causing the callipyge muscle hypertrophy phenotype, the only known example of polar overdominance in mammals. *Genome Res* 12:1496–1506.

Georges M. 2001. Recent progress in livestock genomics and potential impact on breeding programs. *Theriogenology* 55:15–21.

Gootwine E. 2001. Genetic and economic analysis of introgression of the B allele of the FecB (Booroola) gene into the Awassi and Assaf dairy breeds. *Livest Prod Sci* 71:49–58.

Gupta PK, Tsuchiya T. 1991. *Chromosome Engineering in Plants: Genetics, Breeding Evolution.* Amsterdam: Elsevier.

Hale EB. 1969. Domestication and the evolution of behavior. In: Hafez ESE, ed. *The Behavior of Domestic Animals.* Baltimore: Williams & Wilkins. Pp. 22–24.

Heyman Y, Vignon X, Chesne P, Le Bourhis D, Marchal J, Renard JP. 1998. Cloning in cattle: From embryo splitting to somatic nuclear transfer. *Reprod Nutr Dev* 38:595–603.

Johnson LA. 2000. Sexing mammalian sperm for production of offspring: The state-of-the-art. *Anim Reprod Sci* 60–61:93–107.

Khan AS, Fiorotto ML, Hill LA, Malone PB, Cummings KK, Parghi D, Schwartz RJ, Smith RG, Draghia-Akli R. 2002. Maternal GHRH plasmid administration changes pituitary cell lineage and improves progeny growth of pigs. *Endocrinology* 143:3561–3567.

Klein TM, Wolf ED, Wu R, Sanford JC. 1987. High velocity microprojectiles for delivering nucleic acids into living cells. *Nature* 327:70–73.

Kondo JK, McKay LL. 1984. Plasmid transformation of *Streptococcus lactis* protoplasts: Optimization and use in molecular cloning. *Appl Environ Microbiol* 48:252–259.

Kozukue N, Misoo S, Yamada T, Kamijima O, Friedman M. 1999. Inheritance of morphological characters and glycoalkaloids in potatoes of somatic hybrids between dihaploid *Solanum acaule* and tetraploid *Solanum tuberosum. J Agric Food Chem* 47(10):4478-4483.

Larkin P, Scowcroft W. 1981. Somaclonal variation: A novel source of variability from cell culture for plant improvement. *Theor Appl Genet* 60:197–204.

Luchansky JB, Muriana PM, Klaenhammer TR. 1988. Application of electroporation for transfer of plasmid DNA to *Lactobacillus, Lactococcus, Leuconostoc, Listeria, Pediococcus, Bacillus, Staphylococcus, Enterococcus,* and *Propionibacterium. Mol Microbiol* 2:637–646.

Lukaszewski AJ. 2004. Chromosome manipulation and crop improvement. *In: Encyclopedia of Plant and Crop Science.* New York: Marcel Dekker.

Majeskie JL. 1996. Status of United States dairy cattle. In: *National Cooperative Dairy Herd Improvement Program, Fact Sheet K-7.* Washington, DC: U.S. Department of Agriculture.

McHughen A. 2000. *Pandora's Picnic Basket: The Potential and Hazards of Genetically Modified Foods*. New York: Oxford University Press.

NAAB (National Association of Animal Breeders). 2000. *Breakthroughs in Biotechnology: Research Equips Producers with an Array of Genetic Improvement Tools*. Online. Available at http://www.naab-css.org/education/biotech.html. Accessed February 17, 2003.

Notter DR. 1999. The importance of genetic diversity in livestock populations of the future. *J Anim Sci* 77:61–69.

NRC (National Research Council). 2002. *Animal Biotechnology: Science-Based Concerns*. Washington, DC: The National Academies Press.

Patnaik RS, Louie S, Gavrilovic V, Perry K, Stemmer WPC, Ryan CM, del Cardayre S. 2002. Genome shuffling of *Lactobacillus* for improved acid tolerance. *Nat Biotechnol* 20:707–712.

Pina A, Calderon IL, Benitex T. 1986. Intergeneric hybrids of *Saccharomyces cerevisiae* and *Zygosaccharomyces fermentati* obtained by protoplast fusion. *Appl Environ Microbiol* 51:995–1003.

Rexroad CE. 1994. Transgenic farm animals. *ILAR J* 36:5–9.

Rowland GG, McHughen AG, Bhatty RS. 1989. Andro flax. *Can J Plant Sci* 69:911–913.

Rowland GG, McHughen AG, Hormis YA, Rashid KY. 2002. CDC Normandy flax. *Can J Plant Sci* 82: 425–426.

Sanford LL, Kowalski SP, Ronning CM, Deahl KL. 1998. Leptines and other glycoalkaloids in tetraploid *Solanum tuberosum* × *Solanum chacoense* F2 hybrid and backcross families. *Am J Potato Res* 75:167–172.

Sears ER. 1956. The transfer of leaf rust from *Ae. umbellulata* to wheat. *Brookhaven Symp Biol* 9:1–22.

Sears ER. 1981. Transfer of alien genetic material to wheat. In: Evans LT, Peacock WJ, eds. *Wheat Science, Today and Tomorrow*. Cambridge: Cambridge University Press. Pp. 75–89.

Sebastian SA, Chaleff RS. 1987. Soybean mutants with increased tolerance for sulfonylurea herbicides. *Crop Sci* 27:948–952.

Seidel GE Jr. 1984. Applications of embryo transfer and related technologies to cattle. *J Dairy Sci* 67:2786–2796.

Steenson LR, Klaenhammer TR. 1987. Conjugal transfer of plasmid DNA between *Streptococci* immobilized in calcium alginate gel beads. *Appl Environ Microbiol* 53:898–900.

Swanson EB, Couman MP, Brown GL, Patel JD, Beversdorf WD. 1988. The characterization of herbicide tolerant plants in *Brassica napus L* after in vitro selection of microspores and protoplasts. *Plant Cell Rep* 2:83–87.

UPOV (International Union for the Protection of New Varieties of Plants). 2002. *General Introduction to the Examination of Distinctness, Uniformity, and Stability, and the Development of Harmonized Descriptions of New Varieties of Plants*. Online. Available at http://www.upov.int/en/publications/tg-rom/tg001/tg_1_3.pdf. Accessed December 12, 2003.

Zimnoch-Guzowska E, Marczewski W, Lebecka R, Flis B, Schafer-Pregl R, Salamini F, Gebhardt C. 2000. QTL analysis of new sources of resistance to *Erwinia carotovora* ssp. *atroseptica* in potato done by AFLP, RFLP, and resistance-gene-like markers. *Crop Sci* 40:1156–1167.

3

Unintended Effects from Breeding

This chapter explores the likelihood of unintended effects from diverse methods of genetic modification of plants and animals (see Operational Definitions in Chapter 1). Specifically, it discusses unexpected outcomes of breeding methods used to develop a food crop or strain and unexpected or unintended effects recorded in the scientific literature. It also includes analyses of methods intentionally used for modifying food sources and comparing the likelihood of unintended changes resulting from the use of genetic engineering versus other methods of genetic modification discussed in Chapter 2.

BACKGROUND

Novel gene combinations arising from the genetic manipulation of existing genes through conventional breeding techniques may introduce unintended and unexpected effects. However, through the breeder's selection process, the genetic lines that express undesirable characteristics are eliminated from further consideration, and only the best lines—those that express desirable characteristics with no additional undesirable agronomic characteristics, such as increased disease susceptibility or poor grain quality—are maintained for possible commercial release. Although plants and animals produced from conventional breeding methods are routinely evaluated for changes in productivity, reproductive efficiency, reactions to disease, and quality characteristics, they are not routinely evaluated for unintended effects at the molecular level. New varieties of food crops, other than those produced using recombinant deoxyribonucleic acid (rDNA) technologies, are rarely subjected to toxicological or other safety assessments (WHO, 2000). Previous National Academies committees have addressed the question of whether unintended effects arising from the use of rDNA-based technologies in food production and the risks potentially associated with them differ in nature and frequency from those associated with non-rDNA-based breeding methods (NRC, 1987, 2000, 2002).

There is a considerable amount of data compiled and available in the scientific literature addressing issues related to genetically modified (GM) and genetically engineered (GE) plants, including health and environmental impacts. In contrast, development of transgenic animals is a relatively new area of biotechnology, and so the amount of data collected and reported for GM animals is less than that for plants. However, the application of genetic modification techniques and the potential for unintended adverse effects are similar for both plants and animals, and much of the information obtained from plants can be applied to questions of concern in the genetic modification of animals.

PLANT BREEDING

Conventional Plant Breeding

Conventional plant production occasionally generates foods with undesirable traits, some of which are potentially hazardous to human health. Most crops naturally produce allergens, toxins, or other antinutritional substances (see Chapter 5). Standard practice among plant breeders and agronomists includes monitoring the levels of potentially hazardous antinutritional substances relevant to the crop. For example, canola breeders monitor levels of glucosinolates in breeding lines under consideration for prospective commercial release, while potato breeders monitor for glycoalkaloid content. If a particular breeding line generates too much of an undesirable substance, that line is eliminated from consideration for commercial release.

In the United States, the plant breeding community is largely self-monitored. Regulatory agencies do not evaluate conventional new crop varieties for health and environmental safety prior to commercial release. Some other countries require government agencies to conduct premarket evaluations for new crop varieties, both conventional and biotechnology-derived. Canada has a "merit system" for the commercial release of new varieties of major field crops, in which candidate varieties are grown in government-administered field trials. Performance data from these trials, related to agronomic factors, disease resistance, and food quality characteristics, are compiled for all candidate and standard commercial varieties. The data are collected from multiple locations over multiple years, as cereal chemists analyze the grain for chemical and nutritional composition and plant pathologists conduct tests to determine reactions to relevant diseases. These data are then evaluated by a team of experts from industry, government, and universities.

The breeder of each candidate variety must convince these experts that it is competitive and worthy relative to other commercially available varieties of that crop, based on the performance data from the trials. Only if the committee agrees is the candidate variety allowed to be registered as a new commercial variety. If the variety does not perform within prescribed parameters for all characteristics, it is not commercially released (CFIA, 2003).

Unintended Effects of Conventional Plant Breeding

Naturally Occurring Toxins

All foods, whether or not they are genetically engineered, carry potentially hazardous substances or pathogenic microbes and must be properly and prudently assessed to ensure a reasonable degree of safety. Furthermore, all crop strains, including organic strains, potentially express traits generated by various forms of induced mutagenesis. (Under organic regulations, radiation breeding and induced mutagenesis are acceptable, but irradiation of the final food itself is not. For more information on organic regulations, see USDA-AMS, 2001.)

History provides examples of traditional breeding that resulted in potentially hazardous foods (see Box 3-1). Solanaceous (tobacco family) crops, such as potato and tomato, naturally produce various steroidal glycoalkaloids. These substances are toxic not only to humans, but also to insects and pathogenic fungi. During the course of ordinary plant breeding assessments, breeding lines with increased levels of glycoalkaloids may be identified by the breeder as showing superior insect or disease resistance and retained for possible commercial release. The elevation of glycoalkaloid levels responsible for the pest tolerance may not be noted until people become ill from consuming the foods.

Tomatine, a glycoalkaloid naturally present in tomatoes, can be produced in hazardous quantities in certain conventionally bred varieties. Ordinarily, alpha tomatine is present in immature tomato fruit, but is degraded as the fruit matures, so that by the time the fruit ripens to the preferred stage for human consumption, tomatine content is reduced to safe levels. Nevertheless, levels of naturally occurring tomatine in ordinary tomatoes bred using conventional methods can vary considerably, primarily based on maturity, type, and environmental growing conditions (Gilbert and Mohankumaran, 1969; Leonardi et al., 2000). In this respect, environment is more responsible for food hazards than genetic makeup or breeding method.

Another example of a possible effect is the unintended elevation of glycoalkaloid content in potatoes. All potatoes produce the toxic glycoalkaloid solanine, but mature potatoes from most cultivars have amounts so small as to be nonhazardous. However, some varieties produce more than others, and certain environmental stimuli, such as growing or storage conditions, can cause potentially hazardous increases in solanine content, even within a usually safe cultivar (Concon, 1988). For example, potatoes exposed to sunlight turn green, making them particularly prone to high solanine content. Dark-skinned varieties are less

---

BOX 3-1   Kiwi: A New Food?

Puffer fish, chile peppers, and mustard all are traditional foods that humans have learned to consume in moderation to minimize adverse health effects. But what about foods entirely new to the human digestive tract? Although some have designated (GE) foods as completely novel, the GE varieties currently approved and on the market are of the same composition as other foods. Corn oil, for example, is chemically identical regardless of the breeding method used to develop the corn variety.

In recent history, the closest example of a new food might be kiwi fruit. Originally it was an edible but unpalatable plant producing small, hard berries in China. Breeders in New Zealand developed what we know now as kiwi fruit (*Actinidia deliciosa*) into a food during the twentieth century, and commercialized in the United States during the 1960s. There does not, however, appear to be any official record of a premarket safety analysis of the fruit. As a consequence, some humans who were not previously exposed to kiwi fruit developed allergic reactions. Recently, well after commercial release, the responsible allergenic protein (actinidin) was isolated and characterized (Pastorello et al., 1998).

---

likely to turn green than light-skinned varieties, but in either case environment seems to be more important to solanine production than genetics.

Certain potato lines have been found to express greater disease- or pest-resistance, and they have been selected as superior, not always with favorable or intended results. The most notorious such selection was the Lenape potato, which was developed using conventional breeding methods (Akeley et al., 1968). After a successful commercial launch, it was found to have dangerously elevated solanine content in the tubers and was removed from the market (Zitnak and Johnston, 1970). More recently, a similar high-solanine potato variety was detected and withdrawn from the market in Sweden (Hellanäs et al., 1995). In this case, the potato was a heritage variety, developed in the United Kingdom during the nineteenth century, but superseded by another variety due to its susceptibility to disease (Kuiper, 2003). Nevertheless, it became popular in Sweden under the name "Magnum Bonum," until its predilection for overexpressing solanine resulted in its commercial demise (Hellanäs et al., 1995).

In spite of occasional problems with the consumption of potato glycoalkaloids, conventional breeders continue to increase the glycoalkaloid content in the leaves to take advantage of its pest- and pathogen-deterrent properties. Consequently, the U.S. Department of Agriculture (USDA) recommends, but does not require, a limit for glycoalkaloid content in new potato varieties (Sinden and Webb, 1972).

Interestingly, the Lenape (also known as breeding line no. B5141-6) potato has been found to express some useful attributes, such as high solids content. Thus Lenape continues to be used successfully as a parent in conventional breeding programs, providing its genes to new commercial potato varieties, such as Atlantic and Denali. Progeny with genes providing the high solids content are selected or maintained, and the genes responsible for high solanine content in the tubers are selected against or rejected. Additionally, Lenape has been transformed with a genetic construct containing the solanidine glucose-adenosine diphosphate glucosyltransferase (SGT) transgene in the antisense direction, which is designed to interfere with the solanine biosynthesis (Moehs et al., 1997). In field trials, several transgenic Lenape-derived lines expressed substantially less solanine than the parent Lenape, apparently due to antisense expression, an rDNA method for "turning off" an undesirable gene (McCue et al., 2003).

Several breeding programs are developing potatoes derived from conventional crosses between the ordinary potato *Solanum tuberosum* and relatives of other species, such as *S. acaule* (Kozukue et al., 1999) or *S. chacoense* (Sanford et al., 1998; Zimnoch-Guzowska et al., 2000). These are conventional breeding programs in which genes from two different species are exchanged. The intent of these conventional breeding programs is to generate potato varieties with new beneficial features, while minimizing deleterious traits from the foreign species.

Breeders typically monitor levels of toxins in plants that are known to naturally contain them, even though such monitoring is only voluntary in the United States. However, an unexpected and unintended problem may result when combining different species because thousands of genes would be interacting, not just one or two genetic elements of interest. For example, hybrids of *S. tuberosum* and *S. brevidens* produced not only the usual glycoalkaloids, but also the toxin demissidine, which is not produced in either parent (Laurila et al., 1996). This singular result shows that non-genetic engineering breeding methods can have unintended effects and generate potentially hazardous new products.

Any time genes are mutated or combined, as occurs in almost all breeding methods, the possibility of producing a new, potentially hazardous substance exists. Conceivably, similar outcomes could result from using rDNA to transfer specific genes from *S. brevidans* to *S. tuberosum*, giving rise to hybrids expressing the novel toxin demissidine. In either case, the hazard lies with the presence of the toxin, and not with the method of breeding. Genetic engineering could also be used to transfer only the beneficial genes from *S. brevidans*, leaving behind the genes responsible for the novel toxin.

Another example of a toxic compound from traditional crops is psoralens in celery (see Box 3-2). Celery naturally produces these irritant chemicals that deter insects from feeding on the plant and also confer protection from some diseases (Beier and Oertli, 1983). Celery plants with an elevated expression of psoralens will suffer less damage from disease and insect predation and have more aesthetic appeal to consumers, who tend to reject insect- or disease-damaged produce.

**BOX 3-2   Unintended Health Effects:
A Conventionally-Bred Celery Cultivar**

Genetic engineering has raised the question of whether its products create unintended health effects for consumers, but how do conventionally bred products compare along the continuum of genetic modification? Celery provides an interesting case, as it contains naturally occurring toxic chemicals called *psoralens*, which are secondary metabolites found in a variety of fruits and vegetables, and members of the class of compounds called *furanocoumarins* (Diawara and Kulkosky, 2003). The psoralen in celery provides it with a biological defense mechanism of sorts—if it suffers disease or has been bruised, the celery plant can produce up to 100 times the level of psoralen that it typically contains, thus protecting itself from attacks by pests. The production of psoralens also depends on factors such as temperature, season and, particularly, availability of sunlight (Diawara et al., 1995). Psoralens temporarily sensitize human skin to long-wave ultraviolet radiation. Consequently, they have been used for more than 25 years as a component in phototherapy—a means of treating acute skin diseases (New Zealand Dermatological Society, 2002).

However, psoralen in celery has been implicated in cases of skin irritation, such as dermatitis, among farm workers and other handlers of these plants, and studies have suggested a potential correlation between psoralens and cancer in laboratory mice (Beier, 1990). Moreover, celery cultivars produced using conventional breeding methods—intended to enhance insect-resistance and aesthetic appeal to consumers through increased production of psoralen—have been associated with cases of dermatitis among grocery workers, as well as further cases of photosensitivity among farm workers handling these plants (Ames and Gold, 1999). This form of dermatitis, or "photodermatitis," has been observed as far back as 1961 in field workers who had handled celery infected with the disease pink rot (Birmingham et al., 1961). The Centers for Disease Control and Prevention reports cases dating back to 1984 of severe skin rashes among laborers who handled celery on a regular basis. The initial onset of these cases spawned a study by the National Institute for Occupational Safety and Health, in which there appeared to be a relationship between the handling of celery and prolonged exposure to sunlight. In the study that was undertaken in 1984, a number of grocery workers who handled celery on a regular basis also had used a tanning salon, suggesting to researchers that the ultraviolet light from tanning had exacerbated the reactions of the workers' skin in response to the celery they had handled (CDC, 1985). Exposure to elevated levels of psoralens remains a problem for laborers such as field workers, as constant exposure to sunlight—combined with the handling of vegetables such as celery—can induce reactions such as those occurring in the skin. Psoralens continue to be regarded as naturally occurring toxicants (Beier, 1990).

As a result, breeders may observe healthy, undamaged celery lines and select them for commercial release. Unfortunately, workers who harvest high psoralen-producing celery or pack it in grocery stores have, on occasion, developed severe photodermatitis, especially when they are exposed to bright sunlight and the celery is infected with pathogens (Berkley et al., 1986; Birmingham et al., 1961; Finkelstein et al., 1994). There are differences in psoralen content from one variety to another (Beier, 1990; Diawara et al., 1993), but environment seems to have the greatest influence on psoralen production (e.g., Diawara et al., 1995).

Mutations

Mutations, defined as any change in the base sequence of DNA, can either occur spontaneously or be induced, and both methods have produced new crop varieties. Most mutations are deleterious and therefore useless for breeding purposes. However, a mutation can result in desirable traits and may be selected for breeding. Spontaneous mutations, also called "sports" by horticulturists, are by definition not induced, so breeders wait a long time to see mutants arising from this process, and even longer to see useful ones. Most mutation breeders induce random changes in DNA by using ionizing radiation or mutagenic chemicals, such as ethyl methane sulfonate, to increase the rate and frequency of the mutation process.

In spite of these intrusive methods, induced mutagenesis is considered a conventional breeding technique. Food derived from mutation breeding varieties is widely used and accepted. Organic farming systems permit food from mutated varieties to be sold as organic. In the United States many varieties have been developed using induced mutagenesis, such as lettuce, beans, grapefruit, rice, oats, and wheat. The Food and Agriculture Organization of the United Nations/International Atomic Energy Agency Mutant Cultivar Database (FAO/IAEA, 2001) lists more than 2,200 varieties of various species worldwide that have been developed using induced mutagenesis agents, including ionizing irradiation and ethyl methane sulfonate. However, the database does not include spontaneous mutations, cell-selected mutants (Rowland et al., 1989), or somaclonal variants (Rowland et al., 2002).

There is no mandated requirement to list new mutant varieties with the database. However, there do not appear to be outstanding examples of mutant varieties with documented unexpected effects beyond what the mutant was selected for, despite the expectation that mutant varieties may possess and generate more unexpected outcomes than ordinary crosses because of the unpredictable and uncontrollable nature of nontargeted mutations. Furthermore, there do not appear to be any examples in which mutant varieties were removed from the market due to unintended or unexpected adverse incidents.

Genetic Engineering in Plants

As for conventional breeding techniques, the genetic engineering (GE) breeding and selection process is largely self-monitored and only new varieties of rDNA crops are subjected to assessment prior to commercialization. Evaluation of GE food products for other countries is described above, in the section Conventional Plant Breeding.

Genetic engineering methods are considered by some to be more precise than conventional breeding methods because only known and precisely characterized genes are transferred. In contrast, conventional breeding involves transferring thousands of unknown genes with unknown function along with the desired genes. Similarly, in mutation breeding hundreds or thousands of random mutations are induced in each mutated line.

Plant breeders take a variety of methods used to introduce desired traits into consideration during selection. The location where DNA expressing the desired traits is inserted into the host genome in rDNA technology may be immaterial— for example, if an insertion is made in an inappropriate place, the transformed plant is eliminated or selected against by the breeder, who will then select another plant with a preferable locus of insertion for continued evaluation and development as a new variety.

The number of different lines developed by breeders varies according to the crop, the desired trait, and the choice of the breeder, but is it not unusual to start with evaluations of 2,000 "sister" lines from a cross of two parents to develop just one new variety. Thus 99 percent of sister lines are eliminated over the several years of evaluation, for various reasons. These include poor expression of the desired trait, poor yield performance, increased disease susceptibility, or even a lack of visual and tactile appeal, which is a largely subjective and arbitrary designation, but an important criterion nevertheless.

Genetic engineering techniques require fewer lines or transformation events because the desired trait is known and identified early. Consequently, the evaluations—which still take several years—focus on eliminating any unstable lines or those with deleterious characteristics.

In contrast to breeders using other techniques, genetic engineering breeders typically prefer to start with a small number of plants and then select only a few. For example, the two currently approved GE plant varieties from public institutions, papaya (Gonsalves, 1998; Swain and Powell, 2001) and flax (McHughen et al., 1997), started with only about 30 sister lines.

Unintended Effects from Genetic Engineering

Unexpected and unintended effects can be seen with all methods of breeding. Traditionally breeders observe such off-types regularly; they methodologically eliminate these individuals during the evaluation process, long before preparations are made for commercial release. An unexpected or unintended effect does not imply a health hazard, although clearly a plant expressing novel and unexpected characteristics warrants closer inspection prior to commercial release.

A large body of transgenic plant material is developed by university and government laboratories investigating aspects of rDNA that are not related to commercial interests. The Organization for Economic Cooperation and Development (OECD) lists more than 10,000 field trials with GE plants that were conducted between 1986 and 1999. Most of these authorized trials were conducted on plants genetically engineered by public universities and government research institutions, not for commercial release but to test the plants for unexpected or unintended results (see OECD, 2000). Consequently, most of the information on unexpected or unintended results comes from these sources.

Because GE crops are regulated to a greater degree than are conventionally bred, non-GE crops, it is more likely that traits with potentially hazardous characteristics will not pass early developmental phases. For the same reason, it is also more likely that unintentional, potentially hazardous changes will be noticed before commercialization either by the breeding institution or by governmental regulatory agencies.

The following are among the examples of unexpected or unintended characteristics that are often cited in the scientific literature.

Mycotoxin Content in Bt Corn

Bt (*Bacillus thuringiensis*) corn was developed to help farmers protect corn crops from insect pests, particularly the European corn borer and, more recently, the corn root worm. An unexpected effect in this product was a substantial reduction in mycotoxins, which can adversely affect animal and human health. Researchers hypothesized that this is because mycotoxin production is stimulated by fungal spores that infect corn when worms bore into and injure the plant. In the absence of insect damage (as in the *Bt* corn plants), there is reduced opportunity for pathogenic fungal spores to infect the plant and therefore less opportunity to generate the associated and undesirable mycotoxins (Munkvold et al., 1997, 1999).

Increased Lignin in Bt Corn

Saxena and Stotzky (2001) reported that three *Bt* corn varieties increased their stem lignin content relative to their respective non-*Bt* isogenic parents. Since the increase was found in more than a single variety, this suggests that the lignin increase is directly associated with the inserted DNA and is not simply an isolated coincidence. However, increased lignin content has not been recorded in other *Bt* corn lines, so it is uncertain whether the reported increase was due to the rDNA insertion, the presence of *Bt* endotoxin, or some other mechanism. Although lignin is a normal component of plants and of the human diet, and the increased lignin content in the stems of these corn plants is not so great as to present a novel health hazard, the authors do suggest that there might be an environmental consequence from a corn plant with lignin content that is significantly higher than normal.

Petunias with Diminishing Color Over Generations

In this example, a GE petunia line was noticed to have a diminished flower color pattern and intensity over the course of several generations. Molecular analysis showed that a mutation occurred in a gene that regulates flower pigmentation and methylation was responsible for inactivating the gene (Meyer et al., 1992). This phenomenon occasionally occurs with conventional breeding, so its occurrence in a GE petunia does not indicate a unique GE phenomenon. This mutation appears to have occurred in only one GE petunia line out of hundreds generated and tested, thus it is not a general or systematic event.

Insertion of transfer DNA does not guarantee active expression of the associated gene. Inserted genes can be and are silenced or inactivated by the host plant by any of several mechanisms, including methylation (Dominguez et al., 2002). This is a well-known phenomenon and part of the reason genetic engineers generate as many different transgenic plants as possible.

Other Examples

Unexpected or unintentional effects, such as GE soybean stem splitting (Gertz et al., 1999) or *Arabidopsis* with unusual fertility or outcrossing characteristics (Bergelson et al., 1998), have been attributed to the rDNA breeding process.

The GE soybean stem-splitting case documented that a high proportion of GE soybeans suffered split stems in dry, hot conditions. Unfortunately, the non-GE parent line was not measured for stem splitting under similar conditions, but other non-GE soybean varieties were, revealing that a high proportion also suffered split stems. They were not as prone to splitting, however, as the indicated GE line, but that line was within the general accepted range for soybeans. Whether the GE soybean lines tested were more prone to splitting than their parents remains unknown, and thus conclusions cannot be drawn as to whether the genetic engineering process contributed to this trait.

In the *Arabidopsis* example, the authors compared the outcrossing of GE varieties with the outcrossing of a mutated strain that possibly had compromised fertility to begin with and was not an isogenic or near-isogenic line, necessary for properly conducting the type of experiment described.

Other unexpected effects in GE plants have been documented. In GE high oleic soybean lines, metabolic analysis revealed trace amounts of an unintended metabolite, *cis*9,*cis*15-octadecadienoic acid, an isomer of the fatty acid linoleic acid that is not usually present in nonhydrogenated soybean oil, but is present in hydrogenated soybean oil and in other food sources. Because it is a component of other foods, it was not considered a health threat. The developer of the lines, citing Kitamura (1995), argued that the fatty acid substitution was actually nutritionally advantageous. After consideration, the Food and Drug Administration (FDA) determined that the presence of the unexpected linoleic acid isomer and glycinin did not pose a health hazard (DHHS, 1996). USDA also evaluated the GE soybeans, as it does all GE plants, and approved it (APHIS, 1997). FDA did determine that the high oleic soybeans were not substantially equivalent to regular soybeans due to the intended effect of an elevated oleic acid content, thus the breeder had to distinguish them from standard commodity soybeans.

The unexpected effects in these GE soybeans were small and required highly sensitive analytical tools to identify. Such small changes should be expected— although the exact metabolites and levels would remain unpredictable. Furthermore, because genetic changes result in protein changes, metabolic differences due to the presence or actions of new or differing levels of proteins cannot be unexpected. Such changes remain unintended, but they are common occurrences in new crop varieties, including new varieties from conventional breeding (although stringent and detailed analyses of conventional varieties are not conducted).

An important consideration is how these examples of apparent *unpredictability* with rDNA compare with similar unpredicted effects from conventional breeding methods. Unfortunately, the comparisons typically are made only between these unusual rDNA examples and, if conventional plants are compared at all, it is not with isogenic lines but with mutant lines or with less closely related commercial varieties. The correct comparison should be between commercial cultivars of rDNA derivation and isogenic parental varieties. For example, the comparison between *Arabidopsis* and a herbicide-tolerant variety is misleading because the comparator was a mutant variety of uncertain genetic composition. In addition, as discussed earlier, undesirable, unexpected, and unintended traits are noted in conventional breeding lines of commercial crop species on rare occasions, and these lines are consequently discarded.

ANIMAL BREEDING

Conventional Animal Breeding

Genetics is crucial for any livestock enterprise and requires breeders to evaluate production conditions and market objectives; the best breeds or lines of livestock for meeting market goals; selection of a breeding system and breed types; and selection of individual animals within the breed type. Generally, production conditions—so called gene-environment interactions—and market factors have the greatest influence on conventional animal breeding strategies.

Breeders develop a breeding plan that may either be continuous if replace-

---

BOX 3-3   Desirable Traits in Beef Cattle

Certain variables are used in a selection index to make decisions on which sires and dams to use in a breeding plan.
Desirable traits include:

• Environmental adaptability, soundness, temperament, reproduction, livability, longevity, maternal qualities, body size, rate and efficiency of weight gain, and carcass merit.

Additional traits of merit may include:

• Physical size and structure, scrotal circumference as an index of sperm production, semen quality, calving ease, birth weight, weaning weight, average daily weight gain, efficiency of converting feed to body weight gain, weight at 1 year of age, frame score, daughter's maternal ability, fleshing ability, cutability of carcass, intramuscular fat distribution or marbling, and mature size.

---

ment females are selected from within the herd or flock, or be terminal if all breeding females are selected and brought into the herd or flock. The breeding system may involve straight-breeding using only one breed, or cross-breeding using two or more breeds, with the latter expected to provide hybrid vigor, known as heterosis. In sound breeding programs, sires are chosen carefully as they have a great impact on production traits, and dams are chosen carefully based on expected high fertility and desired maternal traits (see Box 3-3). Selection of individual animals for breeding systems within herds or flocks is important, but most animal breeding plans depend on information called "expected progeny difference," which is an estimate from across all animals within a breed of the genetic potential for desired traits an individual can transmit to progeny.

Unintended Effects from Conventional Animal Breeding

Double-Muscling

In beef cattle some non-GE mutant animals show the presence of extraordinary quantities of muscle. The muscular hypertrophy, referred to as double-muscling, is a heritable trait that primarily results from an increase in the number of muscle fibers or cells rather than an increase in the size of individual muscle fibers.

This phenotype has been observed in several cattle breeds. The breed in which this phenotype has been most studied is the Belgian Blue, which has been extensively and systematically selected for double muscling to the point that it is a heritable trait in many herds. Cattle with this phenotype are very lean and produce about 20 percent more lean edible meat (Hanset, 1986; Shahin and Berg, 1985).

Problems with stress tolerance, fertility, and calf viability may have unintended effects in exploiting the double-muscling phenotype for beef production. Researchers established that a genetic deletion in the myostatin gene, which codes for proteins in muscle, was the cause of double muscling in Belgian Blue cattle (Grobet et al. 1997). They also determined the entire coding sequence of the myostatin gene in 32 animals with extreme muscle development from 10 European cattle breeds and found 5 sequence polymorphisms that disrupted the function of the myostatin protein (Grobet et al., 1998).

Porcine Stress Syndrome

In pigs a condition referred to as porcine stress syndrome (PSS), or malignant hyperthermia, has been identified. Affected animals have a higher incidence of death from stress. Pigs with PSS that are harvested at packing plants have pork that is pale in color, exudes excessive water, and is soft in texture. Postmortem, pork from PSS pigs is referred to as PSE—pale, soft, and exudative. PSE pork has significant adverse meat quality attributes, and many consumers find the product objectionable. PSS is the result of intensive breeding of pigs to select animals with increased muscle mass, and it results from an unintended mutation in the ryanodine-receptor gene (Wendt et al., 2000).

Homozygous pigs are more susceptible to stress than are heterozygous pigs. A test that accurately identifies the genetic mutation in pigs that causes PSS has been developed and is widely used in selection programs. Use of this test has resulted in a reduced incidence of the mutation in breeding stock.

Biotechnology

Biotechnology will influence the development of conventional animal breeding. Importantly, those traits that breeders desire to include in their breeding plans are physiologically complex and controlled by multiple genes with variable effects. The overlay of biotechnology and conventional breeding will likely occur in phases that involve broad genetic maps with informative microsatellite markers and evolutionarily conserved gene markers; use of microsatellite markers to identify quantitative trait loci (QTL) of commercially important traits, based on knowledge of complex pedigrees or crosses between phenotypically and genetically divergent breeds or strains; identification of specific trait genes and/or use of conserved markers to identify candidate genes based on their position in gene-rich species, breeds, or strains; and phenotype mapping through functional analysis of trait genes, linking the genome through physiology to the trait.

The use of biotechnology techniques, such as marker-assisted selection, is expected to increase among cattle and swine industries, whereas the poultry industry prefers quantitative genetics. The rigorous use of quantitative genetics in the poultry industry will likely continue as a result of the low cost per animal, short generation intervals, and the ability to address a complex array of traits. Nevertheless, it seems clear that desired complex traits will be understood through the use of genomics information and that the credibility of the technologies of marker assisted and QTL selection in animal breeding programs will be validated and used extensively.

Genetic Engineering

As with plant breeding, animal breeding programs and the production of transgenic farm animals can result in unintended effects. Most, but probably not all, gene-based modifications of animals for food production or therapeutic claims fall under the FDA Center for Veterinary Medicine regulations of new animal drugs. The animal drug provisions of the Federal Food, Drug, and Cosmetic Act provide the legal framework for developing science-based guidelines for assessing the commercial value of these products to society. Other products of transgenic methods will no doubt be developed that could be viewed as containing food or color additives and vaccines. Development of site-specific gene insertion techniques and animal genome projects could change the scope of potential genetic modifications to yield a wider variety of products than are currently being investigated. The following are some examples of unintended effects of recombinant technology administered to animals.

Recombinant Bovine Somatotropin

Commercial sales of recombinant-derived bovine somatotropin (bST) began in the United States during 1994, and its use has gradually increased such that approximately half of U.S. dairy herds, comprising more than 3 million cows, are receiving bST supplements (Bauman, 1999; Etherton and Bauman, 1998). The supplements, which are delivered through an injection of sustained-release formulation every 14 days, result in marked improvements in productive efficiency while maintaining normal cow health and herd life (Bauman, 1999; Etherton and Bauman, 1998). Supplements containing bST are being used commercially in 19 countries.

In animals treated with bST, the yield of milk is increased by 10 to 15 percent (approximately 4-6 kg/d), although even larger increases may occur when the management and care of the animals are excellent (Bauman, 1992; Chilliard, 1989; NRC, 1994). The pattern of response is one in which milk yield gradually increases over the first few days of bST treatment and reaches a maximum during the first week. When treatment is continued, the increased milk yield is sustained. Thus bST results in a greater peak milk yield and an increased persistency in yield over the lactation cycle.

As a consequence of these changes in the lactation curve, commercial practice has shifted to an extended calving interval, which results in fewer births per herd, lower incidence of postpartum metabolic diseases, lower veterinary costs, and an overall improvement in herd life, animal well-being, and dairy farm profitability (Van Amburgh et al., 1997).

Recombinant Porcine Somatotropin

During the past 20 years much has been learned about how porcine somatotropin (pST) increases the growth of pigs (Etherton, 2000; Etherton and Bauman, 1998). These advances were facilitated by the development of methods to produce recombinant-derived pST on a large scale. The availability of large quantities of recombinant pST enabled landmark studies to be conducted that evaluated how the administration of pST affected muscle and adipose tissue growth. Administering pST by intramuscular injection to growing pigs can increase muscle growth by as much as 50 percent and, concurrently, decrease adipose tissue accretion with a maximal effect of about 70 percent. The results of pST administration are due to an array of biological effects from the hormone involving coordinated changes in lipid, protein, and carbohydrate metabolism (Etherton and Bauman, 1998; NRC, 1994). Currently approved for use in 14 countries, pST is undergoing testing required by FDA for commercial use in the United States.

There have been no reported unintended effects at doses of pST that are used in commercial swine production. There is, however, some evidence at much larger doses that pST can cause osteochondrosis, which is an abnormality characterized by an unmineralized, nonvascularized plug of cartilage in the metaphysis of the epiphyseal growth plate (Evock et al., 1988).

Unintended Effects from Genetic Engineering

The report *Animal Biotechnology: Science-Based Concerns* describes some examples of unintended effects found in transgenic pigs (NRC, 2002). The introduction of DNA into random sites in the genome is a mutagenic event that will affect any gene at or near the site of introduction, potentially resulting in unintended effects in the target animal. Given the remarkable variation in sites of gene insertion, the number of gene copies transferred, and the level of gene expression, every animal produced by microinjection is potentially unique in its phenotype (NRC, 2002). This variability contributes enormously to any effort to assess the unintended effects of genetic modification and to evaluate whether any unintended effect that occurs is possibly a cause for a health concern in humans.

An example of a detrimental effect resulting from overexpression of a transgene is illustrated by studies conducted by USDA with transgenic pigs. These studies were the logical extension of earlier work showing that daily injection of recombinant pST markedly increased growth rate, improved the ratio of muscle to fat, and improved feed efficiency (reviewed in Etherton, 2000; Etherton and Bauman, 1998). In these studies, transgenic pigs were created with a gene for human somatotropin or bST to test the hypothesis that overexpression of human somatotropin would result in effects comparable with those observed in studies in which pigs were injected with recombinant pST. Some of the transgenic pig lines did exhibit increased weight gain and were more efficient in converting feed to body weight gain; however, the pigs that showed high levels of either human somatotropin or bST also had a variety of physical problems, including lameness, lethargy, and gastric ulcers (Pursel et al., 1990).

Cloned Animals

A detailed discussion of animal cloning, including the committee's findings and recommendations related to this area, can be found in the subreport included with this volume. The following is a brief overview of unintended effects associated with cloning technology.

Cloning by nuclear transfer from embryonic blastomeres (Willadsen, 1989; Willadsen and Polge, 1981) or from a differentiated cell of an adult (Kuhholzer and Prather, 2000; Polejaeva et al., 2000; Wilmut et al., 1997) requires that the introduced nucleus be reprogrammed by the cytoplasm of the egg and direct development of a new embryo, which is then transferred to a recipient mother to develop to term. The resulting offspring will be identical to their siblings and to the original donor animal in terms of their nuclear DNA, but will differ in their mitochondrial genes and possibly the manner in which the nuclear genes are expressed. Cloning from blastomeres and somatic cells may result in large calves and lambs, the so-called "large offspring syndrome" (Sinclair et al., 2000; Young et al., 1998). More serious abnormalities may also be associated with cloning from somatic cells (Wilmut et al., 1997).

Marker-Assisted Selection

The use of marker-assisted selection in both plants and animals is expected to increase exponentially as data from genome sequencing projects and the density of useful segregating markers increase for economically important species. Marker-assisted selection is already helping plant breeders to identify desirable individuals from heterogeneous populations and to segregate for disease resistance and other features that would ordinarily require several years and substantial field plot experiments. Animals in breeding programs initially will be screened for genes that control simple traits, such as horns, which are undesirable in cattle, and halothane sensitivity, which segregates with metabolic stress syndrome in pigs. As this technique develops over time, easily identifiable markers will be chosen that accompany multiple genes that control more complex traits, such as meat tenderness and taste, growth, calf size, and disease resistance.

Although marker-assisted selection has enormous potential for improving animal health and production traits, it may decrease genetic diversity (Dekkers and Hospital, 2002). Thus short-term gains in productivity may occur at the expense of longer-term improvements due to the extended time before an unintended effect is identified (Dekkers and Hospital, 2002; Dekkers and Van Arendonk, 1998).

It is possible that marker-assisted selection in animal breeding is beneficial because there is a cumulative effect of expression of multiple genes that may be important, whereas effects of individual genes within the chromosomal region associated with the marker are too minor to be of economic importance. Thus the cumulative effects of multiple genes may be exploited to a greater extent by marker-assisted selection for a single gene with a major effect in an attempt to enhance a desired trait. Additionally, this type of marker-assisted selection may counter inbreeding by encouraging breeding strategies that maintain diversity among major genetic loci through exploitation of genes from rare breeds and wild ancestors to improve traits for disease and parasite resistance and adaptability.

MECHANISMS BY WHICH UNINTENDED EFFECTS IN GENETICALLY ENGINEERED ORGANISMS ARISE

Unintended effects may arise from one or more of several mechanisms, which can be either systemic or individual. Systemic effects are those likely to appear in all or almost all transgenic plants, animals, or microbes transformed using the same genetic engineering method and DNA construct. Individual effects, in contrast, are likely due to the nature of a particular transformation event in a single organism, occur in only one transgenic line, and are not displayed by other plants of the same species transformed using the same genetic engineering method. Consequently, individual events may provide greater opportunity to broaden understanding of otherwise unidentified mechanisms that may be responsible for unintended changes.

Some possible mechanisms that may result in unintended effects include the following.

• The sequence of interrupted DNA may be a functional gene, resulting in a loss or gain of whatever function that gene provided.
• Chromosomal changes may occur depending on the location of the insertion.
• The host recipient may have an unusual genotype within the population being sampled that will be expressed and observed in the transgenic organism, an effect known as residual heterozygosity.

• Somaclonal variation or spontaneous mutation may occur in the tissue culture phase of the transformation regeneration processes.

• Interaction may occur between novel gene products and endogenous products—a systemic effect shown in most or all transformation events with the same gene.

When the unexpected effect is systemic and observed with most or all transgenic events, then the likely explanation resides with the function of the expressed introduced gene. For example, an enzyme introduced into a microbe may interact with new substrates, leading to new products in addition to or instead of intended products. That is, the inserted gene might generate the expected protein, but that novel protein might react with metabolites in the new host that differ from those in its native organism. This has occurred in conventional plant breeding when new metabolites, novel to both parents, were generated from crossing *S. tuberosum* and *S. brevidens* (Laurila et al., 1996). Kuiper and colleagues (2001) list several examples of documented unexpected effects in transgenic plants.

Genetic Instability

A frequently stated concern about the use of rDNA technology is that transgenic organisms may be genetically unstable. That is, due to the transfer of genetic material using rDNA methods, the host organism may have some as yet unknown and unobserved mechanism to identify and eliminate the intruding DNA. However, transgenic plants, animals, and microbes have been studied for long enough and intensively enough to determined that genetic instability is not a routine phenomenon. While it is known that transgenic organisms can, on occasion, silence the expression of inserted genes through mechanisms that are now well studied, this phenomenon occurs on an event-by-event basis—not to all transgenic organisms resulting from a particular method or a particular construct.

Although genetically unstable plant varieties can be produced, their likelihood for commercialization is very low. First, if the new trait is unstable, then the organism loses the new trait and reverts back to the traits inherited from the parent plant. Second, domestic and international plant protection regulations exist to protect the intellectual property rights of plant breeders (for more information, see the U.S. Plant Variety Protection Act [P.L. 91-577, 1970]; and the International Convention for the Protection of New Varieties of Plants [UPOV, 1961]). These regulations provide patent-like protection for new varieties of plants under a certificate program. Since only stable varieties are eligible for certification, their production will be favored over unstable varieties.

Gene Transfer Using Foreign DNA

Genetic engineering can be used to transfer DNA within species or across species. In considering examples of cross-species transfers, plants, animals, and microbes all have been engineered with DNA from distant species to give rise to new GE products with realized or potential commercial application. Most examples of commercial GE crops carry DNA from unrelated species, commonly bacteria.

While the process of crossing plant and bacterial DNA is a challenge for conventional plant breeders, gene transfer from bacteria to plants does occur in nature. An example is *Agrobacterium* (see Chapter 2), which has the capability of transferring a portion of its own DNA into plants, where it integrates into the plant genome. In most cases bacteria do not exchange DNA with plants or animals. Genetic engineering transfer of foreign DNA into animals, such as inserting a growth hormone gene into salmon to generate a faster-growing fish, is currently under review by FDA.

Many microbes have been engineered with genes from plants or animals. GE microbes are used to produce a wide range of pharmaceuticals. For example, bacteria are engineered with the human gene for insulin to produce recombinant human insulin. GE microbes are also used to produce products used in food processing. For example, GE bacteria produce the enzyme chymosin used in cheese making. This product was one of the first GE foods approved for use in the United States and in the United Kingdom (FDA, 2001; University of Reading, 2004).

Gene Transfer Using Same-Species DNA

GE foods have been identified as potentially hazardous because they carry genes from foreign species. However, combining genes from different species may or may not be hazardous, as all organisms, including humans, carry genes inserted from different species. For example, all humans carry genes that have been incorporated from viral infections (Cohen and Larsson, 1988).

Genetic engineering is one method that can be used to transfer genes from either related or unrelated species into a host genome. For example, the Xa21 gene is derived from the rice relative *Oryza longistaminata*. This gene confers the valuable agronomic trait of disease resistance, and so various rice varieties have been developed using both conventional crossing and rDNA technology to add the Xa21 gene to commercial rice varieties. Regardless of the method of breeding, from the standpoint of composition, the two new rice varieties are virtually indistinguishable.

Another example includes high oleic acid soybeans, discussed earlier. Both conventional breeding via induced mutagenesis and rDNA transfer of genes from other soybean varieties have been used to generate new soybeans with a higher proportion of oleic acid. However, the conventionally bred high oleic soybeans can be unstable—depending on environmental conditions, their fatty acid content may vary—while the GE high oleic soybean remain more stable (DHHS, 1996).

Other rDNA examples exist where foreign DNA is not necessarily present. The first GE food product approved for human consumption in the United States, the Flavr Savr tomato, was engineered with an inverted tomato gene (CFSAN, 1994), along with bacterial genes transferred to facilitate selection of the modified plant. In the absence of foreign DNA or genes from another species, an argument may be made that a GE plant is not transgenic and not unnatural if only native DNA is transferred. One could readily transfer genetic material from one tomato variety to another, or rice to rice, or soybean to soybean without any other contributing source of DNA.

A recent report of a similar case involved a coffee variety being engineered to produce less caffeine as an alternative to current industrial methods of decaffeination (Ogita et al., 2003), which may use benzene or other organic solvents for extraction. Using rDNA methods, Ogita and colleagues (2003) were able to transform coffee and observe a 70 percent reduction in caffeine in leaves. Because their intent was to interfere with the natural caffeine biosynthetic pathway in the coffee plant, the objective also could have been pursued with an induced mutation breeding program. However, the rDNA method was preferred because it was more precise and predictable, as well as less likely to induce deleterious mutations unrelated to caffeine.

Naturally Occurring rDNA and Human-Mediated rDNA

Genetic recombination occurs in both nature and in human-mediated genetic engineering of plants, animals, and microorganisms (see Box 3-4). This section examines recombination as it occurs in nature and as it occurs with genetic engineering, with particular reference to differences between the two processes. Natural recombination occurs in several ways, typically divided between homologous and nonhomologous recombination events. Genetic recombination events that occur specifically in animals are summarized in Chapter 2 and are discussed in detail in the report *Animal Biotechnology: Science-Based Concerns* (NRC, 2002).

Homologous Recombination

Homologous recombination occurs when DNA strands with similar or homologous base sequences spontaneously recombine with one another. Such juxtaposed sequences do not have to be identical for recombination to occur, but the incidence of recombination between nonidentical sequences diminishes in proportion to the degree of sequence dissimilarity (Lewin, 1985).

Chiasmata are chromosomal crossover events, visible under the microscope, in which whole segments of chromosomes are exchanged. These exchanges often

BOX 3-4 Evolutionary History of Naturally Occurring Hybridization in Plants and Animals

Prior to the use of recombinant deoxyribonucleic acid (rDNA) techniques to achieve recombination of genes from different species, crossing among similar species occurred both with and without human intervention. An example of such naturally occurring "intergeneric" crossing is the wheat that arose thousands of years ago from the incorporation of genes from three different plant species. Many common varieties of modern bread-type wheat carry a fragment of a rye chromosome. When this intergeneric hybridization occurred, it provided the host wheat with a disease-resistance gene—along with an undetermined, uncharacterized, and unregulated number of other genes carried along on the same rye chromosome fragment—and consequently was adopted by plant breeders as a useful source for disease resistance in the wheat gene pool. Such naturally occurring gene transfers across species are frequent; for example, several genetic translocations between wheat and rye have occurred as a result of separate incidents involving commercial bread-type wheat.

Not all species can exchange DNA without human intervention. Even in the case of closely related species, exchanges of genes occur at low frequency and depend on humans to identify and select the rare events to stabilize the genes and traits. Without such intervention, the resulting hybrids would likely die out or take evolutionary time spans to become established, as in the case of bread-type wheat.

Human plant breeders have used their knowledge of interspecies gene transfer to create new species, employing techniques as simple as pollen to pistil transfer. Triticale is an example of an intergeneric hybrid between wheat (*Triticum*) and rye (*Secale*). Triticale was developed as a crop variety that combined the bread-making quality of wheat with the ability of rye to grow in harsh environments (Larter, 1995). Unexpectedly, initial results from this cross expressed the wrong combination that resulted in a weak plant with poor baking quality. Subsequent crosses produced *Triticale* hybrids that worked well enough to be grown with modest commercial success in various regions, including Europe, North America, and Australia. Today, *Triticale* is widely grown and used both in animal feeds and as an ingredient in multigrain breads.

The potato is another example of a common food developed from a combination of genes from different species. Several popular potato varieties were developed using conventional breeding methods to bring useful genes from foreign species into the common potato variety *S. tuberosum* (Jansky and Rouse, 2000). Additionally, some *Nicotiana* species carry genes naturally transferred by the bacterium, *Agrobacterium*, through nonhomologous recombination (Aoki et al., 1994; Frundt et al., 1998). While these naturally occurring events are possible either with or without human intervention, genetic engineering allows for the introduction of a greater variety of novel genes into a host genome (see Figure 3-1).

involve dozens or even hundreds of genes, giving rise to new chromosomal structures. Such naturally occurring events are crucial to conventional crossbreeding as they allow the breakage of chromosomally linked undesirable genes from useful genes and the recombination of desirable genes.

Nonhomologous Recombination

Nonhomologous recombination is observed in species from phage and bacteria to plants and animals. As noted in Chapter 2, transposable elements—also known as transposons—are well-characterized genetic segments with the ability to insert into nonhomologous DNA sequences and to copy themselves and move elsewhere. Results of transposon activity include stable integrations into the target DNA and loss of genetic function if the insertion interrupts a previously functional gene; complete excisions with restoration of function of previously interrupted gene; and rearrangements around the site of integration.

Nonhomologous recombination allows the DNA to be inserted into dissimilar sequences, which means almost anywhere in the genome. Such latitude allows the possibility of insertion into a functional gene, thus inactivating it, in contrast to homologous recombination where the DNA sequence at the locus of insertion is known, that is, identical or near identical with the insertional DNA with the recombined gene generally retaining functionality. This factor makes nonhomologous recombination less attractive to breeders than homologous recombination because nonhomologous recombination events will have to be more stringently screened to identify and eliminate inserts into and possible inactivation of nontarget genes.

In practice, insertion of transferred DNA, or T-DNA, into crucial genes is a relatively rare event in crop species and in application to crop improvement. The standard breeding practices of candidate variety evaluation and screening to eliminate lines showing undesirable changes identifies and eliminates those lines with inserts that inactivate important genes. Genetic engineering also has the distinct advantage over non-genetic engineering methods because the sequence of the inserted DNA is known, and it can be used to identify the precise location of the insert in the genome and characterize nearby genes.

Preferential Insert Loci

Recombinational "hotspots" are hypothetical locations in the genome where T-DNA is preferentially inserted. The implication is that these targeted loci, for which sound evidence in plants is lacking, are regions of unstable DNA so any insert would be similarly unstable. It seems logical to predict, then, that if an unstable segment of DNA is inserted into an unstable sequence of host DNA, then the combined instability would lead to unstable, unintended results.

This interesting conjecture remained unanswered for years and was the basis of considerable debate in the *Agrobacterium* transformation community. Recent analyses of *Agrobacterium* T-DNA inserts into *Arabidopsis* provide data to address the question. The Salk Institute maintains a population of 32,500 *Arabidopsis thaliana* lines with *Agrobacterium*-mediated T-DNA inserts in and around the 25,500 genes in the genome of this species. An analysis shows the loci of insertion to be reasonably evenly distributed along the entire genome. Some of the inserts are between genes and others interrupt genes (Ecker, 2003).

In a related study comparing T-DNA with transposons inserted into the *Arabidopsis* genome, for transposons a slight preference was found for insertion into coding sequences of genes. As with the study of Ecker (2003), Pan and colleagues (2003) found that both T-DNA and transposon insertions are reasonably evenly distributed throughout the genome.

These data indicate that there is no strong recombinational hotspot or strong preferential insert site for T-DNA. Similarly, there is no evidence to suggest the CaMV 35s promoter in GE plants is any more unstable than the CaMV 35s promoter in ordinary plants infected with CaMV.

Recombination Hotspots and Genetic Instability

The range of loci in the *Arabidopsis* genome into which T-DNA has been inserted indicates that the insertion location is random or near random (Ecker, 2003). If insertional hotspots do exist, they clearly are not exclusive, but at most they are mildly preferred points of recombinational insertion. Genetic instability at singular loci in itself is not hazardous unless the unstable nature necessarily gives rise to hazardous outcomes. However, genetic instability, when it occurs, affects only one cell or plant at a time, and the result is typically a loss of activity of the relevant gene. An example is the case where a herbicide tolerance gene is inserted into an unstable locus and then grown in a field of the herbicide-tolerant plants. When an instability incident occurs in this crop, the herbicide tolerance is lost—the plants experiencing this instability event revert to being herbicide susceptible. When the field is then sprayed with the herbicide, only those few individual plants suffering the instability succumb to the spray.

Because the timing of such destabilizing events is rare and does not strike an entire population simultaneously, it is unlikely that the loss of a small number of crop plants would be noticed compared with the loss of weeds or with nonherbicide-tolerant volunteer crop plants. If genetic instability were more common or synchronized, such that it were noticeable, the breeder would not be able to obtain certification for the new variety due to failure to meet the genetic stability requirement.

Genetic instability is observed from time to time, both in conventional breeding and in genetic engineering. Robertson (1978) described a conventional corn line with a high degree of genetic instability due to an endogenous transposon. As discussed earlier, unstable pigmentation in GE petunias provides an example of a systemic instability in particular GE plants (Meyer et al., 1992).

Spontaneous mutations change DNA also, affecting everything from single base changes, known as "point mutations," to entire genomes. Deletions of entire segments from a genome are predictably severe and deleterious, but point mutations or even more substantial mutations, such as those affecting large tracts of DNA in chromosomes, can be beneficial and adaptive.

Because of the ubiquitous nature of spontaneous mutations, all plant varieties—whether conventional or GE—exhibit a small degree of genetic instability and generate mutants with some degree of frequency. If the frequency of spontaneous genetic instability is low, the typical loss-of-function mutants will probably not be noticed; if it is high, the variety cannot be commercialized.

Natural types of recombination can also result in the same effects. Transposons, even *Agrobacterium* insertions, all interrupt any DNA sequence where they insert, regardless of whether they insert into homologous or nonhomologous DNA. Depending on the function of the interrupted DNA, there may or may not be phenotypic consequences from the insertion.

Agrobacterium naturally inserts DNA into the host plant-cell genome, typically in a genetically stable manner. The substitution of desirable DNA by genetic engineering for the phyto-oncogenic DNA of the wild strains does not affect the mechanics of transfer. The genes responsible for *Agrobacterium* genetic transformation and recombination in the plant are physically and functionally separate from the T-DNA actually transferred and integrated. The mechanics of *Agrobacterium* appear the same whether the event is staged through a natural infection process by a crown gall-producing wild-type strain or by the same strain that has been disarmed by the removal of the gall-producing genes and subsequently replaced with known, desirable DNA.

THE GENETIC MANIPULATION CONTINUUM

Overview

Predicting the likelihood of unintended hazards from compositional changes associated with genetic modifications does not fit a simple dichotomy comparing genetic engineering with non-genetic engineering breeding. This is because there are many mechanisms shared in common by both GE and non-GE methods, and also because there are techniques that slightly overlap each other. Furthermore, within the scope of genetic engineering (rDNA) technology, several mechanisms for genetically transforming plants are available as options to scientists, such as *Agrobacterium*-mediated gene transfer and the use of the particle gun (McHughen, 2000). These two examples of genetic engineering are as different mechanistically as the conventional methods of narrow crossing and wide crossing (discussed in Chapter 2). Consequently, it is unlikely that all methods of either genetic engineering or conventional breeding will have equal probability of resulting in unintended changes. It is the final product of a given modification, rather than the modification method or process itself, that is more likely to result in an unintended adverse health effect. In Figure 3-1 some of the methods used to generate GM plants are shown to illustrate the full range of possibilities that might lead to unintended changes (these methods are described in Chapter 2).

Analysis of the Continuum

As noted earlier in this chapter, unintentional changes are possible with all conventional and biotechnological breeding methods for genetic modification. It is possible to represent the likelihood of such changes as a continuum, albeit only a partially understood one. Placement along this continuum has no bearing on risk of adverse outcomes, but only on the probability of unintended changes, which need not be hazardous. The potential for hazard resides in specific products of the modification regardless of whether the modification was intentional or unintentional.

Again, unintended effects do not necessarily imply hazard. If a particular method were inherently hazardous, all products resulting from its use would be potentially harmful. However, it is known that each method can provide safe products, so the key for breeders and regulatory agencies, in their reviews of specific products, is to identify the relatively rare, potentially hazardous products resulting from any method.

In breeding, the developer can generate literally thousands of breeding lines; each might be progeny of the same parents, thus bringing together the same set of genetic information, and yet possess subtle differences. As noted previously in this chapter, conventional breeders may scrutinize 2,000 of these "sister lines" of a single cross of two parent plants. Meiotic recombination allows sister lines to show a continuum of phenotypic expression. Over several years, the breeder eliminates most, or sometimes all, of those breeding lines because they are unsuitable.

Plant breeding is often said to be a process not of selection, but of elimination. Any off-types, unstable lines, or lines showing characteristics such as significant differences in nutrient content, responses to environmental stresses, diseases, or the presence of other undesirable traits are discarded as soon as they are noticed. This winnowing takes place over several years, so the remaining lines identified for prospective commercial release are unlikely, but not guaranteed, to have any significant compositional changes other than those related to the desired trait. For this reason, regulatory scrutiny focuses most often on the new trait and its metabolic perturbations. Nevertheless, the appearance of subtle or obscure phenotypic changes can go unnoticed by breeders or regulators and may subsequently have to be recalled from the market, as in the case of the Lenape potato—

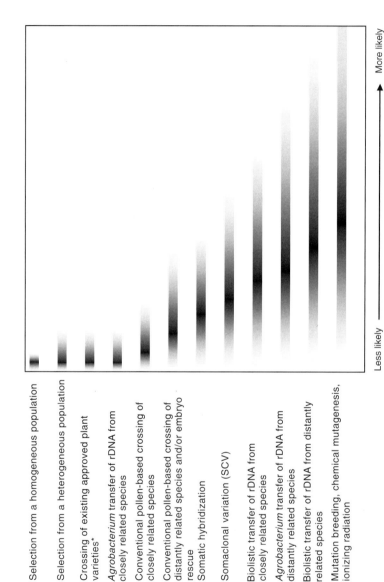

Selection from a homogeneous population

Selection from a heterogeneous population

Crossing of existing approved plant varieties*

Agrobacterium transfer of rDNA from closely related species

Conventional pollen-based crossing of closely related species

Conventional pollen-based crossing of distantly related species and/or embryo rescue

Somatic hybridization

Somaclonal variation (SCV)

Biolistic transfer of rDNA from closely related species

Agrobacterium transfer of rDNA from distantly related species

Biolistic transfer of rDNA from distantly related species

Mutation breeding, chemical mutagenesis, ionizing radiation

Less likely   More likely

*includes all methods of breeding

developed using non-genetic engineering methods—discussed previously in this chapter (Laurila et al., 1996).

The products of a narrow cross-pollination, from a plant of a given species to another plant of the same species, or even the same subspecies, are less likely, but not guaranteed, to pose unintended risks. This is due to the expectation that pollination is inherently safer than other methods of gene transfer because the range of resulting products is typically limited to those products already present in the species or subspecies. However, even this likely "safe" method is subject to occasional spontaneous mutations or infelicitous genetic recombination events leading to novel and unexpected products, and these products may carry some degree of risk, such as disease susceptibility, leading to, for example, hazardous levels of fungal mycotoxins. Expanding the range of conventional breeding beyond the same species provides a greater opportunity to introduce novel genetic information and thereby novel sources for unexpected effects.

Triticale is a man-made crop developed by breeding wheat with rye; *hordecale* is a similar product of barley and rye. Both examples bring together thousands of genes in a way that does not occur in nature and which conceivably could result in the production of undesirable toxicants. As discussed previously in this chapter, the conventional potato breeding program combining *Solanum tuberosum* and *S. brevidens* produced not only the expected toxicants, but also a new one, demissidine, which is not produced in either parent (Laurila et al., 1996).

Although some products can be reasonably predicted to be hazardous, this does not imply that all products of rDNA—or other methods of bringing together genes—are necessarily hazardous. *Agrobacterium* and the particle gun are the only two rDNA methods represented on the continuum in Figure 3-1 because they provide almost all of the GE crops approved for commercial release. They are depicted separately because of the documented predilection of the particle gun to introduce multiple, broken, or rearranged DNA segments, in contrast with the usually singular and high-fidelity transfer typical of *Agrobacterium.*

FIGURE 3-1 Relative likelihood of unintended genetic effects associated with various methods of plant genetic modification. The gray tails indicate the committee's conclusions about the relative degree of the range of potential unintended changes; the dark bars indicate the relative degree of genetic disruption for each method. It is unlikely that all methods of either genetic engineering, genetic modification, or conventional breeding will have equal probability of resulting in unintended changes. Therefore, it is the final product of a given modification, rather than the modification method or process, that is more likely to result in an unintended adverse effect. For example, of the methods shown, a selection from a homogenous population is least likely to express unintended effects, and the range of those that do appear is quite limited. In contrast, induced mutagenesis is the most genetically disruptive and, consequently, most likely to display unintended effects from the widest potential range of phenotypic effects.

The more disorganized particle gun mechanism increases the risk of insertion into crucial endogenous sequences. The degree of increased risk is as yet undetermined and may be minimal. However, in both cases sufficient numbers of events are generated and screened to make it possible to identify those with the most desirable insertion features, the best expression of desired genes, and the least likelihood of deleterious features, in much the same way that conventional breeders evaluate their breeding lines.

Continuum Conclusions

Unintended adverse health effects can be either predictable or unpredictable. All organisms undergo spontaneous mutations, giving rise to novel traits, which may carry some hazard. The incidence of such mutations is relatively rare and the type of hazard associated with them is generally not predictable. On the other hand, introduced mutations, such as by rDNA, theoretically allow a gene from any species to be inserted into and expressed by a food crop. Clearly, such a product has the potential to be hazardous if the inserted gene results in the production of a hazardous substance. Although the process of rDNA is itself not inherently hazardous, the resulting product of the process may be.

DISCUSSION

This chapter reviews examples of unintentional changes that resulted from the genetic modification of various organisms intended for food and the likelihood of unintentional changes arising from multiple methods of genetic modification. All forms of genetic modification, conventional and modern, may potentially lead to unintended changes in composition, some of which may have adverse health effects.

The range of biotechnology methods, including genetic engineering, makes it possible to alter, add, or remove genes from conventional food organisms. This manipulation of genetic information may pose either risks or benefits to health or the environment. Examples abound in which a single gene can have a dramatic effect or no effect at all, depending on its features. Similarly, chromosomal changes can also have either a dramatic or no apparent effect, again depending on the features of the genes involved (NRC, 2002).

Risks to human health from genetic changes in foods must, therefore, be placed in proper context. The introduction of allergenic proteins is a potential adverse health effect of concern that could arise from genetic modification, including genetic engineering, of food. These methods of breeding, however, are not the only way that potential allergens are introduced into the food supply (see Chapter 5).

Although genetic modification techniques may introduce unpredicted adverse health effects, a given technique itself is not a determinant of many more common adverse effects, such as those that result from personal food intolerances or sensitivities; crops growing in soils containing toxicants; contamination from postharvest infection with pathogens; and adulterants inadvertently mixed into food. By comparison, genetic modifications to food account for a lesser likelihood for adverse health effects.

Most of the novel features in commercially available GE products focus on single genes, such as herbicide tolerance or pest resistance in crops. Products developed using biotechnology express a wide range of new features, some of which may be benign, while others, such as industrial or pharmaceutical products, may pose a greater threat to food safety. Conventional breeding methods have also provided new crop varieties with enhanced herbicide tolerance, pest resistance, and other non-nutrient characteristics, as well as enhanced nutritional profiles, posing potential risks from ingestion that are similar to genetic engineering. Similar comparative examples can be found in animal breeding and in microbial strains, in which conventional methods have produced new products with unintended hazards similar to those that may result as a consequence of the application of rDNA technology.

REFERENCES

Akeley RV, Mills WR, Cunningham CE, Watts J. 1968. Lenape: A new potato variety high in solids and chipping quality. *Am Potato J* 45:142-145.

Ames BN, Gold LS. 1999. Pollution, pesticides and cancer misconceptions. Pp. 18-39 in Fearing Food, J. Morris and R. Bate, eds. Oxford, UK: Butterworth Heinemann.

Aoki S, Kawaoka A, Sekine M, Ichikawa T, Fujita T, Shinmyo A, Syono K. 1994. Sequence of the cellular T-DNA in the untransformed genome of *Nicotiana glauca* that is homologous to ORFs 13 and 14 of the Ri plasmid and analysis of its expression in genetic tumors of *N. glauca* × *N. langsdorfi*. *Mol Gen Genet* 243:706–710.

APHIS (Animal and Plant Health Inspection Service). 1997. *DuPont Petition 97-008-01p for Determination of Nonregulated Status for Transgenic High Oleic Acid Soybean Sublines G94-1, G94-19, and G-168. Environmental Assessment and Finding of No Significant Impact.* Online. U.S. Department of Agriculture. Online. Available at http://www.aphis.usda.gov/brs/dec_docs/9700801p_ea.HTM. Accessed December 12, 2002.

Bauman DE. 1992. Bovine somatotropin: Review of an emerging animal technology. *J Dairy Sci* 75:3432–3451.

Bauman DE. 1999. Bovine somatotropin and lactation: From basic science to commercial application. *Domest Anim Endocrinol* 17:101–116.

Beier RC.1990. Natural pesticides and bioactive components in foods. *Rev Environ Contam Toxicol* 113:47-137.

Beier RC, Oertli EH. 1983. Psoralen and other linear furocoumarins as phytoalexins in celery. *Phytochemistry* 22:2595–2597.

Bergelson J, Purrington CB, Wichmann G. 1998. Promiscuity in transgenic plants. *Nature* 395:25.

Berkley SF, Hightower AW, Beier RC, Fleming DW, Brokopp CD, Ivie GW, Broome CV. 1986. Dermatitis in grocery workers associated with high natural concentrations of furanocoumarins in celery. *Ann Intern Med* 105:351-355.

Birmingham, DJ, Key MM, Tubich GE, Perone VB. 1961. Phototoxic bullae among celery harvesters. *Arch Dermatol* 83:127-41.

CDC (Centers for Disease Control and Prevention). 1985. Phytophotodermatitis among grocery workers. Morbidity and Mortality Weekly Report 34(1):11–13. Online. Available at http://www.cdc.gov/mmwr/preview/mmwrhtml/00000464.htm. Accessed August 26, 2003.

CFIA (Canadian Food Inspection Agency). 2003. *Seeds Act (R.S. 1985, c. S-8): Seeds Regulations. Updated to April 30, 2003.* Online. Department of Justice, Canada. Available at http://laws.justice.gc.ca./en/S-8/C.R.C.-c.1400/172987.html. Accessed August 21, 2003.

CFSAN (Center for Food Safety and Applied Nutrition). 1994. *FDA Backgrounder: Biotechnology of Food.* Online. Food and Drug Administration. Online. Available at http://www.cfsan.fda.gov/~lrd/biotechn.html. Accessed December 12, 2002.

Chilliard Y. 1989. Long-term effects of recombinant bovine somatotropin (rBST) on dairy cow performances: A review. In: Sejrsen K, Vestergaard M, Neimann-Sorensen A, eds. *Use of Somatotropin in Livestock Production.* New York: Elsevier Applied Science. Pp. 61–87.

Cohen M, Larsson E. 1988. Human endogenous retroviruses. *Bioessays* 9:191–196.

Concon J. 1988. Food toxicology: Principles and concepts. Part A. New York, N.Y.: Marcel Dekker, Inc.

Dekkers JC, Hospital F. 2002. The use of molecular genetics in the improvement of agricultural populations. *Nat Rev Genet* 3:22–32.

Dekkers JCM, van Arendonk JAM. 1998. Optimum selection for quantitative traits with information on an identified locus in outbred populations. *Genet Res* 71:257–275.

DHHS (U.S. Department of Health and Human Services). 1996. *High Oleic Acid Transgenic Soybean.* Online. Memorandum of Conference. Online. Available at http://www.cfsan.fda.gov/~acrobat2/bnfM039.pdf. Accessed November 14, 2002.

Diawara MM, Kulkosky PJ. 2003. Reproductive toxicity of the psoralens. *Pediatr Pathol Mol Med* 22(3): 247–259.

Diawara MM, Trumble JT, Quiros CF. 1993. Linear furanocoumarins of three celery breeding lines: implications for integrated pest management. *J Agric Food Chem* 41:819-824.

Diawara MM, Trumble JT, Quiros CF, Hansen R. 1995. Implications of distribution of linear furanocoumarins within celery. *J Agric Food Chem* 43:723–727.

Dominguez A, Fagoaga C, Navarro L, Moreno P, Pena L. 2002. Regeneration of transgenic citrus plants under nonselective conditions results in high-frequency recovery of plants with silenced transgenes. *Mol Genet Genomics* 267:544–546.

Ecker J. 2003. *A Sequence-Indexed Library of Insertion Mutations in the Arabidopsis Genome.* Online. Salk Institute Genomic Analysis Laboratory. Available at http://signal.salk.edu/tabout.html. Accessed October 14, 2003.

Etherton TD. 2000. The biology of somatotropin in adipose tissue growth and nutrient partitioning. *J Nutr* 130:2623–2625.

Etherton TD, Bauman DE. 1998. The biology of somatotropin in growth and lactation of domestic animals. *Physiol Rev* 78:745–761.

Evock CM, Etherton TD, Chung CS, Ivy RE. 1988. Pituitary porcine growth hormone (pGH) and a recombinant pGH analog stimulate pig growth performance in a similar manner. *J Anim Sci* 66:1928–1941.

FAO/IAEA (Food and Agriculture Organization of the United Nations/International Atomic Energy Agency). 2001. *FAO/IAEA Mutant Varieties Database.* Online. Available at http://www-infocris.iaea.org/MVD/. Accessed January 1, 2003.

Finkelstein A, Afek U, Gross E, Aharoni N, Rosenberg L, Halevy S. 1994. An outbreak of phytophotodermatitis due to celery. *Int J Dermatol* 33:116-118.

FDA (U.S. Food and Drug Administration). 2001. Partial list of microorganisms and microbial-derived ingredients that are used in foods. Washington, DC: U.S. Food and Drug Administration, Center for Food Safety and Applied Nutrition. Available at http://vm.cfsan.fda.gov/~dms/opamicr.html. Online. Accessed December 3, 2003.

Frundt C, Meyer AD, Ichikawa T, Meins F. 1998. A tobacco homologue of the Ri-plasmid orf13 gene causes cell proliferation in carrot root discs. *Mol Gen Genet* 259:559–568.

Gertz JM, Vencill WK, Hill NS. 1999. Tolerance of transgenic soybean (*Glycine max*) to heat stress. In: *Proceedings of the 1999 Brighton Conference—Weeds*. Farnham, UK: British Crop Protection Council. Pp. 835–840.

Gilbert JC, Mohankumaran N. 1969. High tomatine tomato breeding lines. *Veg Impr Newslett* 11:6.

Gonsalves D. 1998. Control of papaya ringspot virus in papaya: A case study. *Annu Rev Phytopathol* 36:415–437.

Grobet L, Royo Martin LJ, Poncelet D, Pirottin D, Brouwers B, Riquet J, Schoerberlein A, Dunner S, Menissier F, Massabanda J, Fries R, Hanset R, Georges M. 1997. A deletion in the myostatin gene causes double-muscling in cattle. *Nat Genet* 17:71–74.

Grobet L, Poncelet D, Royo LJ, Brouwers B, Pirottin D, Michaux C, Menissier F, Zanotti M, Dunner S, Georges M. 1998. Molecular definition of an allelic series of mutations disrupting the myostatin function and causing double-muscling in cattle. *Mamm Genom* 9:210–213.

Hanset R. 1986. Double muscling in cattle. In: Smith C, King JW, McKay JW, eds. *Exploiting New Technologies in Animal Breeding: Genetic Development*. Oxford: Oxford University Press. Pp. 71–80.

Hellanäs KE, Branzel C, Johnsson H, Slanina P. 1995. High levels of glycoalkaloids in the established Swedish potato variety Magnum Bonum. *J Sci Food Agric* 68: 249-255.

Jansky SH, Rouse DI. 2000. Identification of potato interspecific hybrids resistant to Verticillium wilt and determination of criteria for resistance assessment. *Potato Res* 43:239–251.

Kitamura K. 1995. Genetic improvement of nutritional and food processing quality in soybean. *Jarq-Jpn Agr Res Q* 29:1–8.

Kozukue N, Misoo S, Yamada T, Kamijima O, Friedman M. 1999. Inheritance of morphological characters and glycoalkaloids in potatoes of somatic hybrids between dihaploid *Solanum acaule* and tetraploid *Solanum tuberosum*. *J Agric Food Chem* 47(10):4478-4483.

Kuhholzer B, Prather RS. 2000. Advances in livestock nuclear transfer. *Proc Soc Exp Biol Med* 224:240–245.

Kuiper HA. 2003. Biotechnology, the environment, and sustainability. *Nutr Rev* 61(6, Supplement 1):105-109.

Kuiper HA, Kleter GA, Noteborn HP, Kok EJ. 2001. Assessment of the food safety issues related to genetically modified foods. *Plant J* 27:503–528.

Larter EN. 1995. Triticale. In: Slinkard AE, Knott DR, eds. *Harvest of Gold. The History of Field Crop Breeding in Canada*. Saskatoon: University of Saskatchewan. P. 367.

Laurila J, Laakso I, Valkonen JPT, Hiltunen R, Pehu E. 1996. Formation of parental type and novel alkaloids in somatic hybrids between *Solanum brevidens* and *S. tuberosum*. *Plant Sci* 118:145–155.

Leonardi C, Ambrosino P, Esposito F, Fogliano V. 2000. Antioxidative activity and carotenoid and tomatine contents in different typologies of fresh consumption tomatoes. *J Agric Food Chem* 48:4723-4727.

Lewin B. 1985. *Genes II*. New York: John Wiley & Sons.

McCue KF, Allen PV, Rockhold DR, Maccree MM, Belknap WR, Shephard LVT, Davies H, Joyce P, Corsini DL, Moehs CP. 2003. Reduction of total steroidal glycoalkaloids in potato tubers using antisense constructs of a gene encoding a solanidine glucosyl transferase. *Acta Hortic* 619:77-86.

McHughen A. 2000. *Pandora's Picnic Basket: The Potential and Hazards of Genetically Modified Food*. New York: Oxford University Press.

McHughen A, Rowland GG, Holm FA, Bhatty RS, Kenaschuk EO. 1997. CDC Triffid transgenic flax. *Can J Plant Sci* 77:641–643.

Meyer PF, Linn F, Heidmann I, Meyer H, Niedenhof I, Saedler H. 1992. Endogenous and environmental factors influence 35s promoter methylation of a maize A1 construct in transgenic petunia and its color phenotype. *Mol Gen Genet* 231:345–352.

Moehs CP, Allen PV, Friedman M, and Belknap WR. 1997. Cloning and expression of solanidine UDP-glucose glucosyltransferase from potato. *Plant J* 11:227-236.

Munkvold GP, Hellmich RL, Showers WB.1997. Reduced *Fusarium* ear rot and symptomless infection in kernel of maize genetically engineered for European corn borer resistance. *Phytopathology* 87:1071–1077.

Munkvold GP, Hellmich RL, Rice LG. 1999. Comparison of fumonisin concentrations in kernels of trangenic *Bt* maize hybrids and nontransgenic hybrids. *Plant Dis* 83:130–138.

New Zealand Dermatological Society. 2002. Patient information: PUVA. Online. Available at http://www.dermnetnz.org/index.html. Accessed August 26, 2003.

NRC (National Research Council). 1987. *Introduction of Recombinant DNA-Engineered Organisms into the Environment: Key Issues*. Washington, DC: National Academy Press.

NRC. 1994. *Metabolic Modifiers: Effects on the Nutrient Requirements of Food-Producing Animals*. Washington, DC: National Academy Press.

NRC. 2000. *Genetically Modified Pest-Protected Plants: Science and Regulation*. Washington, DC: National Academy Press.

NRC. 2002. *Animal Biotechnology: Science-Based Concerns*. Washington, DC: The National Academies Press.

Ogita S, Hirotaka U, Yamaguchi Y, Koizumi N, Sano H. 2003. RNA interference: Producing decaffeinated coffee plants. *Nature* 423:823.

OECD (Organization for Economic Cooperation and Development). 2000. BIOBIN: A co-operative resource on safety in biotechnology, developed between OECD's BioTrack Online and UNIDO's BINAS. Online. Available at http://www1.oecd.org/ehs/biobin/. Accessed October 5, 2003.

Pan X, Liu H, Clarke J, Jones J, Bevan M, Stein L. 2003. ATIDB: *Arabidopsis thaliana* insertion database. *Nucleic Acids Res* 31:1245–1251.

Pastorello EA, Conti A, Pravettoni V, Farioli L, Rivolta F, Ansaloni R, Ispano M, Incorvaia C, Giuffrida MG, Ortolani C. 1998. Identification of actinidin as the major allergen of kiwi fruit. *J Allergy Clin Immunol* 101:531-537.

Polejaeva IA, Chen SH, Vaught TD, Page RL, Mullins J, Ball S, Dai Y, Boone J, Walker S, Ayares DL, Colman A, Campbell KH. 2000. Cloned pigs produced by nuclear transfer from adult somatic cells. *Nature* 407:86–90.

P.L. (Public Law) 91-577. 1970. Plant Variety Protection Act, 84 Stat. 1542—1559. Online. Available at www.ams.usda.gov/science/PVPO/PVPO_Act/whole.pdf. Accessed April 11, 2003.

Pursel VG, Hammer RE, Bolt DJ, Palmiter RD, Brinster RL. 1990. Integration, expression and germline transmission of growth related genes in pigs. *J Reprod Fertil Suppl* 41:77–87.

Robertson DS. 1978. Characterization of a mutator system in maize. *Mutat Res* 51:21–28.

Rowland GG, McHughen A, Bhatty RS. 1989. Andro flax. *Can J Plant Sci* 69:911–913.

Rowland GG, McHughen AG, Hormis YA, Rashid KY. 2002. CDC Normandy flax. *Can J Plant Sci* 82:425–426.

Sanford LL, Kowalski SP, Ronning, CM, Deahl KL. 1998. Leptines and other glycoalkaloids in tetraploid *Solanum tuberosum* × *Solanum chacoense* F2 hybrid and backcross families. *Am J Potato Res* 75:167-172.

Saxena D, Stotzky G. 2001. *Bt* corn has a higher lignin content than non-*Bt* corn. *Am J Bot* 88:1704–1706.

Shahin KA, Berg RT. 1985. Growth patterns of muscle, fat, and bone, and carcass composition of double muscled and normal cattle. *Can J Anim Sci* 65:279–293.

Sinclair KD, Young LE, Wilmut I, McEvoy TG. 2000. In-utero overgrowth in ruminants following embryo culture: Lessons from mice and a warning to men. *Human Reprod* 15:68S–86S.

Sinden SL, Webb RE. 1972. Effect of variety and location on the glycoalkaloid content of potatoes. *Am Potato J* 49:334-338.

Swain S, Powell DA. 2001. *Papaya Ringpot Virus-Resistant Papaya: A Case Study*. Online. University of Guelph. Available at http://www.foodsafetynetwork.ca/gmo/papayarep.htm. Accessed October 10, 2002.

University of Reading. 2004. Genetically modified food case studies: Chymosin. Reading, UK: University of Reading, National Centre for Biotechnology Education. Online. Available at http://www.ncbe.reading.ac.uk/NCBE/GMFOOD/chymosin.html. Accessed January 3, 2004.

UPOV (International Convention for the Protection of New Varieties of Plants). 1961. International Convention for the Protection of New Varieties of Plants, adopted by the Diplomatic Conference on December 1, 1961. Online. Available at http://www.upov.int/en/publications/conventions/1961/content.htm. Accessed May 22, 2003.

USDA-AMS (U.S. Department of Agriculture-Agricultural Marketing Service). 2001. The National Organic Program. Washington, DC: U.S. Department of Agriculture-Agricultural Marketing Service. Online. Available at http://www.ams.usda.gov/nop/indexIE.htm. Accessed June 14, 2003.

Van Amburgh ME, Galton DM, Bauman DE, Everett RW. 1997. Management and economics of extended calving intervals with use of bST. *Livest Prod Sci* 50:15–28.

Wendt M, Bickhardt K, Herzog A, Fischer A, Martens H, Richter T. 2000. Porcine stress syndrome and PSE meat: Clinical symptoms, pathogenesis, etiology and animal rights aspects. *Berl Munch Tierarztl Wochenschr* 113:173–190.

WHO (World Health Organization). 2000. *Safety Aspects of Genetically Modified Foods of Plant Origin*. Report of a joint FAO/WHO Expert Consultation. Geneva: WHO.

Willadsen SM. 1989. Cloning of sheep and cow embryos. *Genome* 31:956–962.

Willadsen SM, Polge C. 1991. Attempts to produce monozygotic quadruplets in cattle by blastomere separation. *Vet Rec* 108:211–213.

Wilmut I, Schnieke AE, McWhir J, Kind AJ, Campbell KH. 1997. Viable offspring derived from fetal and adult mammalian cells. *Nature* 385:810–813.

Young LE, Sinclair KD, Wilmut I. 1998. Large offspring syndrome in cattle and sheep. *Rev Reprod* 3:155–163.

Zimnoch-Guzowska E, Marczewski W, Lebecka R, Flis B, Schafer-Pregl R, Salamini F, Gebhardt C. 2000. QTL analysis of new sources of resistance to *Erwinia carotovora* ssp. *atroseptica* in potato done by AFLP, RFLP, and resistance-gene-like markers. *Crop Sci*, 40:1156-1167.

Zitnak A, Johnston GR. 1970. Glycoalkaloid content of B5141-6 potatoes. *Am Potato J*, 47: 256-260

4

New Approaches for Identifying
Unintended Changes in Food Composition

BACKGROUND

Many advances have occurred in recent years that have extended our ability to determine the chemical composition of food and other biological material with greater depth, accuracy, and precision. Improved analytical methodology has many benefits in the study of food composition, including a far more in-depth understanding of nutrient content, relationships between chemical composition and acceptability (e.g., patterns of flavor compounds), and the safety of the food material.

This improved analytical capability should, in principle, also provide a basis for evaluating the compositional differences between particular foodstuffs as a function of genetic variables, environmental factors, and agricultural practices. It is this comparative approach that is perhaps the most important in the safety evaluation of new food items derived from all means of genetic modification, including conventional breeding techniques, mutagenesis techniques, or genetic engineering (see Operational Definitions in Chapter 1).

Important advances in analytical methodology for nucleic acids, proteins, and small molecules have occurred over the past decade as a result of concurrent advances in technology and instrumentation. Many such techniques are becoming relatively more user-friendly, and instrumentation has become available at lower cost. Consequently, more laboratories have the ability to conduct detailed analyses of food composition. This situation has highlighted the need for validating and standardizing methods, certifying laboratory performance through both time and location, and having reliable standards and certified reference materials that are broadly and uniformly available. There is a great need for improvement in all of these areas, although it is clear that analytical techniques, such as profiling, will continue to develop and improve independent of the need to apply these methods to the assessment of genetically modified (GM) and genetically engineered (GE) foods.

Advanced molecular genetic, proteomic, and metabolite profiling techniques are rapidly developing technologies that have the potential to provide an enormous amount of data for a given organism, tissue, or food product. The levels of analysis include:

- Deoxyribonucleic acid (DNA) sequence analysis (i.e., the complete sequence of an organism's genome or the targeted sequencing of a transgene insertion site to determine whether insertion into the genome is in a location likely to affect the expression of adjacent genes).
- Gene expression analysis to determine alterations in the levels of messenger ribonucleic acid (mRNA) species.
- Protein analysis to determine the pattern, identity, and relative abundance of specific proteins (i.e., Are proteins of catalytic, allergenic, or toxicological concern present?).
- Specific organic compounds, especially small molecules and trace elements whose presence, pattern of relative concentrations, and absolute concentration provide information of nutritional, antinutritional, and toxicological relevance.

In theory, these data sets could be used, singly or in combination, in comparative studies to assess the nutritional quality and chemical composition of food in relationship to the environment, genetics, naturally occurring or induced mutations, and genetic engineering. A possible secondary benefit is the generation of a wealth of data that may ultimately contribute to a better understanding of the fundamental linkages between food composition and health.

An ideal situation for any analytical procedure would provide the following information:

- The absolute structural identification of all compounds in a sample being analyzed.
- The absolute quantification of all compounds, taking into account varying recovery and detection sensitivity for each compound in the sample.
- The biological or biochemical impact of each compound (positive, neutral, or negative) in isolation and in a complex mixture at a given dosage.
- The relative nutritional (or antinutritional) importance in the human diet of a compound from a given food, and the significance of modifying the concentration of this compound, on the overall nutrient profile of the general population.
- The ability to perform predictive modeling of the changes to a target food organism's metabolism and physiology as the result of a transgenic event and predictive modeling of the biological consequences of these changes to human health.

Although these items represent the analytical ideal, in almost all of these instances the current procedures for chemically analyzing food components and assessing the impact of food components on health fall well short of ideal. This chapter discusses various approaches to food analysis involving advanced and emerging analytical methods and their application. The discussions in this chapter are meant to apply to plants, animals, and microbes. Plants are the most frequently cited example because the introduction of transgenic plants into the food supply is much more pervasive and advanced than either animals or microbes. It should be recognized that improvements are occurring rapidly for both targeted and untargeted (i.e., profiling) methods.

TARGETED QUANTITATIVE ANALYSIS VERSUS PROFILING METHODS

Two basic analytical approaches exist, and each has merit in certain applications. Targeted quantitative analysis is the traditional approach in which a method is established to quantify a predefined compound or class of compounds (e.g., amino acids, lipids, vitamins, or RNAs for specific genes). In contrast, profiling methods involve the untargeted analysis of a complex mixture of compounds extracted from a biological sample with the objective of determining the pattern of detected constituents. For proteins and metabolites this is most often accomplished either by chromatographic (e.g., gas chromatography-mass spectrometry [GC-MS] or liquid chromatography-mass spectrometry [LC-MS]), electrophoretic, or spectral (e.g., nuclear magnetic resonance [NMR]) means, while for nucleic acids methods based on sequence-specific hybridization are used.

The ultimate goal in profiling methods is to quantify and identify all compounds present in a sample (i.e., RNA, protein, and metabolites). This goal is closer to being realized for RNA (the expression of genes) due to advances in gene chip technology and the fact that all DNA and RNA are composed of nucleic acids. Complete quantification and identification of all proteins and metabolites in a sample is still only a theoretical possibility for the reasons discussed in the following sections.

Profiling methods in general are intended to determine the relationship between the pattern of components and a quantitative attribute (e.g., as used widely in the sensory analysis field to evaluate compounds associated with desirable flavor or odor attributes) and to identify differences in the composition of samples by comparing chromatographic and/or spectral patterns derived from complex mixtures. A positive characteristic of the profiling methods is the fact that they allow comparison of patterns of constituents and detection of compositional differences without the requirement for identification of all of the compounds or an understanding of the functions of all genes in an organism.

The inherent difficulties, however, in identifying all of the constituents detected in profiling methods or understanding the activity and potential biological consequences of all genes in an organism severely limit the usefulness of these methods for predictive purposes, especially in extremely complex samples, such as most plant and animal tissues and products. In addition, profiling methods are limited by issues of sensitivity, constraints imposed by sample extraction and preparation methods, and possible artifacts generated in sample handling, extraction, and preparation. This directly impacts the ability to assess the biological consequences of any changes (real or artifactual) that are observed. Selected examples of both targeted analysis and profiling methods, which illustrate the advantages and current limitations, are discussed below.

General Considerations for Accuracy, Reproducibility, and Artifacts in Analysis of Food and Other Biological Material

Important characteristics of any analytical technique are the precision and accuracy of the chosen method, the robustness and reproducibility of the method within and between laboratories, and the ongoing identification and appreciation of any potential sources of artifacts in the methods employed. Such issues are often broadly grouped into the categories of technical variation, biological variation, and artifacts. The impact of these different categories can vary greatly among different analytical procedures and, thus, the following guidelines are intended only for general consideration.

Technical variation arises from small differences in the reactivity and stability of a compound and the chemistry and physical properties of the analytical procedure. Technical variation is most often determined as the variation from the mean following the repeated processing and analysis of the same sample. It is generally the lowest source of variation in an analytical method, but it is not identical for all compounds and can vary several-fold for different compounds targeted in an analysis. Technical variation generally ranges from 1 to 20 percent and is due primarily to the differential stability and chemical reactivity of individual compounds during extraction, isolation, separation, and quantification procedures.

Biological variation, in general, is often several-fold higher than technical variation, is independent of the analytical procedure used, and can vary significantly for each compound in an analysis. Biological variation most likely arises from small interindividual differences in the growth and development of organisms and from the interactions of an organism and the environment under apparently identical conditions.

Artifacts are, by their very nature, unpredictable and can arise from alternative reactions with reagents, with or among endogenous sample compounds, or from interactions with components inherent to the analysis (e.g., column matrices or buffer components). Artifacts can generate signals, peaks, or compounds that are not present in the original sample, or they can induce the disappearance or reduction of genuine peaks, signals, or compounds that are present in the original sample. The potential for artifacts in any procedure is an unavoidable consequence of extraction, isolation, chemical modification, and detection.

For targeted analyses (e.g., particular vitamins or amino acids) technical variation and artifacts can often be minimized as only a limited number of compounds of often similar chemistry are being targeted and analyzed. Extraction and manipulation procedures can therefore be optimized to favor the isolation of target compounds over other compounds, while simultaneously excluding many potentially interfering compounds. Similarly, the chemical manipulations required for separation and detection also can be tailored to favor the target compounds, and the potential for artifact generation is often reduced.

Profiling methods often utilize extraction procedures that are not selective and allow a wide range of compounds (with varying chemistries) to be isolated, including any interfering or confounding compounds that may be present. Similarly, the methods of separation and detection are often a compromise to allow for a broad range of compounds to be detected and, hence, the potential for higher degrees of technical variation and artifacts is increased. It should be recognized that biological variation is essentially independent of analytical technique and would be similar for both targeted and profiling techniques.

Considerations and Strategies in Targeted Analysis of Metabolites and Other Constituents

Advances in analytical chemistry have been applied widely to the quantitative analysis of specific compounds and classes of compounds in food and related biological material. For many nutrients, toxic compounds, and other food constituents, modern methodologies allow a more sensitive, precise, and rapid analysis than could previously be accomplished. However, complicated aspects of sample preparation often limit laboratory throughput, and many issues of calibration, standardization, and other quality control considerations remain to be satisfactorily resolved.

In spite of the overall advances in food analysis, a complete analysis of all nutrients and potentially relevant phytochemicals and other compounds of interest remains an arduous task for even the most advanced analytical laboratory because the complete identification and quantification of all compounds in a sample, plant, or product is yet to be accomplished. Thus the development of a paradigm of analytical requirements that focuses on the most important compositional questions would be the most prudent in terms of effective use of analytical resources and consumer safety. It is better to devote analytical resources to a thorough determination of the most nutritionally and toxicologically relevant compounds than to broaden the analysis unnecessarily by including analyses for compounds that have little importance to overall health and food safety. The following discussion illustrates this rationale and includes representative examples. This should not be viewed as an exhaustive list.

There does not appear to be an appropriate one-size-fits-all approach for targeted analysis when evaluating the safety of new food derived from genetic engineering or conventional breeding. However, as stated above, it is reasonable to propose a minimal set of analyses, in addition to basic proximate composition, as a starting point. As discussed below there are several variations on and extensions of current analytical practice that can achieve a more complete understanding of the pattern of nutrients, toxicants, antinutrients, and other relevant constituents. In this framework, requirements for additional specific analyses could be selected on a case-by-case basis according to the particular chemical composition (e.g., nutrient profile) and potential risks associated with a given type of food product.

A realistic goal in the analytical evaluation of GE food is measuring the content of relevant essential micronutrients (vitamins and minerals), essential macronutrients (essential amino acids and fatty acids), nonessential nutrients of importance to health (e.g., dietary fiber, total fat), antinutrients (e.g., enzyme inhibitors and lectins), and known toxicants. With respect to nutritional composition, primary analytical attention should be directed toward those compounds of greatest importance. For example, legumes (such as soybeans) do not constitute a significant source of dietary ascorbic acid, but they can be an important source of folate for some populations. Analytical requirements and priorities should be established accordingly.

The nature of the genetic change introduced also should be considered in establishing analytical requirements. For example, a product in which genetically introduced changes alter total protein content, expression of a specific protein or synthesis of one or more amino acids should trigger a requirement for analysis of the full pattern of amino acids. Similarly, alterations affecting the synthesis of any type or class of lipids (e.g., altered fatty acid profile or altered sterol content) should trigger a requirement for full analysis of a fatty acid profile, as well as other lipid classes and related compounds (e.g., fat-soluble vitamins).

The potential to improve the nutritional quality of a food by genetically upregulating the biosynthesis or storage of nutrients or other plant components already has been demonstrated (Yan and Kerr, 2002), and this continues to be an active area of research and development. Such changes can be achieved through breeding, by genetic engineering of the enzymes of an entire biosynthetic pathway or enzymes that have a high impact on product synthesis or accumulation, or by the biosynthesis of a limiting precursor. Compositional engineering of this type raises the potential for unintended changes in the chemical composition of the resulting food. This could occur through changes in the overall flux of total carbon or alterations of flux through pathways that supply multiple aspects of plant or animal metabolism (e.g., isoprenoid or methyl group synthesis).

There also is the potential that increasing the concentration of a precursor molecule could lead to greater concentrations of that chemical's metabolites. For example, plants often have a large capacity to glycosylate (a biochemical modification to add carbohydrate structures to a molecule) many compounds, and an increase in a precursor molecule that is subject to such glycosylation could lead to an increase in its glycosylated derivatives. In addition, increasing the concentration of a chemical compound, such as a vitamin or other phytochemical, could lead to an increased concentration of catabolic products. While these examples may or may not be relevant to food safety, they illustrate several ways in which unexpected compositional changes could occur.

Nutrients

Several analytical issues should be noted with respect to the measurement of nutrients. More in-depth discussions of nutrients and nutritional considerations are included in Chapters 5 and 6.

Vitamins. Many vitamins exist as a group of structurally related chemical compounds. In some cases, all of the various forms of a vitamin exhibit approximately the same biological activity and bioavailability for humans. However, in other cases, the various forms may have different biological properties. For example, members of the carotenoid family exhibit large differences in vitamin A activity.

In the case of vitamin E, understanding of the vitamin E activity of compounds has changed recently, and this should be reflected in the interpretation of analytical data. Although all members of the tocopherol and tocotrienol families exhibit antioxidant activity, only alpha-tocopherol and its esters appear to exhibit activity in satisfying the human nutritional need for vitamin E (IOM, 2000). Any change in the alpha-tocopherol content, regardless of the relative proportions of other tocopherols or tocotrienols, would alter the net nutritional value of the food product. This principle was demonstrated in an experimental enhancement of tocopherol synthesis that, when accompanied by an engineered increase in the conversion of gamma-tocopherol to the more nutritionally active alpha-tocopherol, led to a large increase in vitamin E activity (Shintani and DellaPenna, 1998).

Another consideration with respect to vitamins and the selection or development of targeted analytical methods is the relationship between chemical form and bioavailability. A classical example is the heterogeneous group of conjugated or "bound" forms of niacin that exist in corn and certain cereal grains. Little is known about the genetic or environmental factors that affect the conversion of niacin compounds to such unavailable forms, but they have been shown to change with maturation in corn (Wall et al., 1987). Measuring free (i.e., bioavailable) niacin, in addition to total niacin, as has been conducted in many classical analyses, should be conducted in grains using contemporary methods specific for the available forms of this vitamin. It should be noted that this issue would be of lesser importance for grain products destined to be used in food or feed in which nutritional supplementation, enrichment, or fortification with niacin is practiced.

The potential for alteration in vitamin bioavailability also exists in many plants for vitamins other than niacin that can undergo glycosylation (Gregory, 1998). For example, a substantial fraction of the vitamin B_6 in many foods from plant origin can exist as a beta-glucoside that exhibits only partial bioavailability in humans. A change in the proportions of free and glycosylated forms of vitamin B_6 could alter the overall nutritional characteristics of a fruit, vegetable, or grain. Thus a focused analysis that included a measurement of all forms of this vitamin, glycosylated and nonglycosylated, or a full assessment of nutritional properties in plant substances that are important sources of this vitamin would be necessary.

Amino Acids. Many advances have occurred in the measurement of the amino acid content of food and other biological material, such as improved resolution, speed, and sensitivity. Genetic engineering and conventional breeding practices have the potential to alter the amino acid content of food either intentionally or unintentionally, and such changes have the potential to be nutritionally important, especially in plant-derived food. The limiting essential amino acid in legume crops is typically methionine, while cereal grains are generally limiting in lysine and/or threonine. Genetically induced changes in protein expression could either lessen or accentuate these nutritional limitations, as could changes in the biosynthesis of the essential amino acids.

Fatty Acids. Changes in the concentration of total fat and the profile of fatty acids in oilseed crops can have significant nutritional effects, and such compositional changes can be mediated through both genetic engineering and nongenetic-based engineering methods (Thelen and Ohlrogge, 2002). GC methods allow measurement of the full distribution of fatty acids and should be included in the focused analysis of oilseed crops and other organisms that contribute significantly to dietary lipid intake. Modern high-resolution GC allows for the detection of a wide range of fatty acids and their geometric and positional isomers. This analysis should include identification, by GC-MS, of novel fatty acids detected in new food being evaluated, whether the food was derived from genetic engineering or from conventional breeding.

Dietary Fiber and Related Constituents. It is not likely that major changes in total dietary fiber content would occur in new plants derived from either genetic engineering or conventional breeding. However, in view of the importance of plants as the primary sources of dietary fiber and the potential for dietary fiber constituents to affect the bioavailability of certain nutrients, measurement of total dietary fiber should be performed. Further analysis to quantify the individual classes of fiber constituents seems unnecessary unless evidence of changes in total dietary fiber is found.

A report of increased lignin in stalks of various lines of GE (*Bt*) corn (Saxena and Stotzky, 2001) suggests that compositional changes in dietary fiber should not be overlooked, although more specific analytical methods, including the use of a rigorous sampling regimen, would be required to allow a more comprehensive interpretation of this reported phenomenon. Measurement of total dietary fiber, as well as the various major dietary fiber constituents, would be necessary. The concentration of other potentially undesirable constituents should also be considered when appropriate. For example, has the concentration of galactosyl sucroses (flatulence factors) been increased in new lines of legumes or certain other plants? Contemporary high-performance LC (HPLC) and LC-MS methods are well suited for such a separation and quantitative analysis of these oligosaccharides.

Food Constituents that Potentially Affect Nutrient Bioavailability

Although the concept of nutrient bioavailability is very complex, several factors that affect bioavailability are sufficiently well-characterized to merit their inclusion in a targeted analysis of relevant food. For example, phytate is a common constituent of cereal grains that affects the bioavailability of many divalent cationic minerals (e.g., zinc and iron). Oxalate is commonly found in green leafy vegetables and affects mineral bioavailability by a mechanism similar to that of phytate. Another example is the digestibility of protein from oilseeds, which is known to be poor in the uncooked state. This is due to the presence of various enzyme inhibitors and lectins, as well as to the natural resistance of native oilseed proteins to digestion. The measurement of these components would be prudent in new lines of soybeans and other constituents.

Biologically Active Non-nutritive Compounds

As stated above, biologically active and potentially toxic compounds should be selected for analysis as appropriate on the basis on their natural existence. The following is a brief discussion of several plant components that should also be considered.

Mycotoxins. The importance of determining secondary toxins from other organisms (e.g., mycotoxins) should not be overlooked. It is possible that compositional and structural changes in plant tissues due to genetic engineering could make the plants either more or less susceptible to mycotoxin contamination as a result of differences in insect resistance and, hence, susceptibility to mold infestation and contamination with mycotoxins (e.g., aflatoxins and fumonisins). For example, several reports indicate that corn engineered to express the *Bacillus thuringiensis* toxin has lower fumonisin content, presumably due to decreased insect herbivory and, hence, decreased introduction of fungi into the plant tissues (Bakan et al., 2002; Clements et al., 2003; Dowd, 2001; Duvik, 2001). In view of the potential for such variability in susceptibility, routine analysis of mycotoxins, as is already routine practice, continues to be warranted. Great advances have been made in the measurement of mycotoxins using immunochemical and traditional HPLC methods.

Phytoestrogens and Other Non-nutritive Bioactive Constituents. Isoflavones and lignans are common constituents of soy and related legumes that exhibit estrogenic activity and have some potential for enhancement through genetic engineering (Liu et al., 2002). Because these compounds may have positive health effects for some consumers and adverse effects when present in excessive quantities, an understanding of phytoestrogen content and bioactivity is important. Thus phytoestrogens should be considered candidates for routine targeted analysis of relevant plants. The results of such analyses should be interpreted with caution, however, because of the potential for large variability in phytoestrogen levels among samples as a function of both genetic (i.e., variety) and environmental factors (Lee et al., 2003).

Hormones. Changes in the levels of a hormone (e.g., as a consequence of biotechnology) may produce a variety of effects, depending on the magnitude of the change. Hormone actions are complicated by several factors. Hormones cannot act unless there is an activated receptor. Thus, if the process of genetic modification or engineering does not impact on expression of the receptor, the likelihood of a metabolic effect would be minimal. Additionally, hormones often have indirect actions on nontarget tissues. For example, stromal cells that express receptors for sex steroids also produce growth factors, which affect overlying epithelial cells. Identification of an unintended adverse effect would require an assessment for effects on multiple cell types in order to detect potential downstream effects and determine whether there is a potential risk to health that may result. Testing criteria for GM, including GE, products with respect to primary hormones and potential downstream effectors may be useful to identify unintended health effects.

Alkaloids. The types of alkaloids present in plant tissue are highly species-dependent (Ashihara and Crozier, 2001; Verpoorte and Memelink, 2002). These include, but may not be limited to, gossypol, tomatine, caffeine, and solanine. Contemporary HPLC, GC, LC-MS, and GC-MS methods facilitate rapid and sensitive targeted analyses of specific alkaloids.

Targeted Analysis of DNA, RNA, and Proteins

To provide a complete picture of the genetic and compositional changes of food produced by either genetic engineering or conventional breeding, a targeted analysis using the tools of modern molecular biology should be used to provide information regarding the specific genetic changes that have occurred.

For example, a Southern blot or similar analysis could be used to confirm the introduction of one or more new genes. Sequencing upstream and downstream from the transgene would provide information regarding the site of the insertion and the possibility that the insertion has disrupted another gene or its regulatory element.

It may also be prudent to conduct a targeted analysis of the transgene transcript (and genes adjacent to the insertion site) by using gene-specific probes and methods such as Northern blot analysis or real time-polymerase chain reaction to verify and quantify the extent, developmental timing, and tissue specificity of transgene expression to ascertain that other adjacent genes are not impacted by their expression in the transgenic line. Perhaps most important in a food safety assessment is the measurement of the expression of the protein encoded by the transgene and, if enzymatically active, the concentration of the reaction products and their metabolites, as discussed above. Information on approaches to safety assessment of GM foods is also available in the Report of the Fourth Session of the Codex Ad Hoc Intergovernmental Task Force on Foods Derived from Biotechnology (FAO/WHO, 2003). The novel protein may also be assessed for allergenic potential and other safety considerations as discussed in Chapter 6.

NONTARGETED ANALYTICAL METHODS FOR METABOLITES

The term metabolomics has been used to describe the nontargeted analysis of small organic molecules in a complex sample. In theory the method should allow unambiguous identification and precise and reproducible quantification and detection of all chemical constituents of a sample, even those varying in concentration by several thousand-fold. In practice the chemical and physical manipulations required for the method, combined with the extreme diversity of the compounds being analyzed, do not allow all compounds to be studied in a single analysis.

The differential requirements for solubility, stability, and detection of different compounds and compound classes, coupled with the current limitations for absolute structural identification, pose significant limitations on the application of this methodology for use in detecting and determining the unintended consequences in a food. Indeed, the complete identification and quantification of all compounds in a sample, plant, or product are still far from a reality, even for the most intensively studied model biological systems (e.g., *E. coli*, *Arabidopsis*, yeast). Additionally, as discussed earlier and below, even if and when an unintended consequence is demonstrated, whether as a result of breeding, chemical mutagenesis, or genetic engineering, it is most often difficult to impossible to predict, based on this information alone, the effect (if any) the change might have on human or animal nutrition, biology, and health, even for extremely well-studied compounds.

Analytical Approaches

Improvements in chromatographic and spectroscopic instrumentation, innovative chromatographic and electrophoretic separation techniques, and enhanced data-processing capabilities have led to major improvements in our understanding of food composition. In particular, separation techniques (e.g., HPLC, GC, and capillary electrophoresis) have greatly extended the capability of analysts to resolve the components of complex mixtures for both small molecules and macromolecules.

Major advances in MS instrumentation have led to the widespread availability of compact, highly sensitive, relatively inexpensive, and user-friendly HPLC-MS and GC-MS equipment suitable for quantitative and qualitative analysis. Such analytical improvements also have led to a better ability to detect and quantify compositional changes associated with biological variation and with variables such as agricultural practices, climate, and genetics. In addition, the application of such techniques has led to the recognition that the composition of biological material, especially plants, is far more complex and variable than previously believed, and that the majority of plant chemical constituents have yet to be identified and structurally characterized.

The potential of these contemporary methods in the detection of compositional changes in plant tissues has been discussed in reviews by Kuiper and associates (2001, 2003). These same techniques are also applicable to the analysis of metabolites in tissues and fluids from humans and test animals, and these may have applicability in the evaluation of the metabolic impact of changes in food composition (German et al., 2003).

Advantages and Disadvantages

It is believed that plants, from a purely biochemical perspective, are among the most chemically complex organisms on the planet. The total number of different compounds produced by all plants is currently estimated to be between 100,000 and 200,000 and is likely to be several times higher as analytical methodologies improve (Fiehn, 2002). The number of compounds produced by a single plant species may vary between 5,000 and 10,000. Given the complexity of metabolites in plants, no single analytical methodology is currently available that will achieve resolution and quantification of all compounds in a plant tissue. Therefore, several different methods are often used, individually or in parallel, to attempt to resolve and quantify, and where possible, to absolutely identify compounds in a plant mixture.

Three of the more common methodologies, GC-MS, LC-MS, and NMR are briefly described below. For all three methods and for the study of plant metabolism in general, major limitations include:

- A lack of universally accepted standardized methods for extraction, separation, and quantification of metabolites
- A lack of spectral libraries that would allow the unambiguous identification of a peak from a given analysis
- A need for improved data management and data-mining systems in order to derive useful information from data sets generated through research.

Another limiting factor is the lack of commercial availability of many known natural products, thus making the task of generating spectral standards for identification more difficult. A consequence of recent consolidation in the preparative chemical industry has been a reduction in the catalog of commercially available chemicals. Efficient application of these analytical techniques may require setting up a resource capable of producing purified natural products either from natural sources or from chemical synthesis.

GC-MS Analysis

One of the most widely used and robust analytical methods for metabolite profiling is GC-MS. Typically plant tissues are extracted with various combinations of organic and aqueous solvent systems. All compounds cannot be extracted by a single solvent system, and usually several systems that differ in polarity are used. The compounds present in each extraction are then analyzed. Derivatization is a requirement for the separation of most compounds by GC, but it introduces the complication that the procedure can modify the target molecule such that absolute structural determination of the original molecule is not possible.

Despite the limitations of differing solubility and chemical derivatization, several studies have shown that 100 to 500 individual compounds can be reproducibly resolved in a single GC-MS analysis of plant tissue extracts. Generally 20 to 40 percent of the molecules can be unambiguously identified based on published mass spectra (Fiehn et al., 2000a, 2000b; Roessner et al., 2000). Although this represents only a fraction of the estimated 5,000 to 10,000 compounds in a given plant tissue, it is nonetheless an important advancement in the ability to broadly characterize metabolites in a nontargeted fashion. The technical variation of most GC-MS methodologies is generally less than 10 percent, while the biological variation encountered in several comprehensive studies averages 50 percent and is highly dependent on the particular compound in question (Fiehn et al., 2000b; Roessner et al., 2000).

GC-MS metabolite profiling has recently been applied to a number of experimental plant systems. In studies of potato tubers (Roessner et al., 2000, 2001), more than 150 compounds were resolved, of which 77 could be identified. Technical variation was 6 percent or lower for 29 of 33 compounds analyzed, while the biological variation for these same compounds ranged from 2 to 36 percent and exceeded technical variation by two- to tenfold. When tubers grown in soil were compared with those grown in sterile culture with an external carbon source (glucose), large differences in a range of metabolites (sugars, amino acids, organic acids, and unknowns) were observed. Similarly, potatoes genetically engineered to have significant changes in primary carbohydrate metabolism (Roessner et al., 2001) showed significant differences in a large range of metabolites (sugars, amino acids, organic acids, and unknowns).

Another series of metabolite profiling studies utilized the model plant system *Arabidopsis thaliana*. In one set of studies (Fiehn et al., 2000a, 2000b), more than 326 distinct compounds were identified, of which the structures of 164 could be unambiguously determined, while 162 were of unknown chemical structure. Technical variation in these studies was less than 10 percent, while biological variation for 11 compounds studied in detail ranged from 17 to 56 percent, with an average of approximately 40 percent.

Two mutants derived from chemical mutagenesis were compared with their respective wild-type parental lines (Fiehn et al., 2000a). The dgd1 mutant caused a 90 percent reduction in the galactolipid digalactosyldiacylgylcerol, a major component of the chloroplast membrane. In this study, 153 of 326 metabolites (known and unknown) changed significantly in the dgd1 mutant compared with the wild type. A second mutant, sdd1, affected the number of stomata on the leaf surface. (Stomata are pairs of cells that work together to regulate gas exchange between the leaf and the atmosphere.) Because stomata are a minor component of the leaf, one might expect the sdd1 mutation to impact metabolism less severely than dgd1. This was indeed found to be the case as the sdd1 mutant had 41 metabolites altered relative to the wild type. It is important to note that neither the sdd1 nor dgd1 mutant is a result of genetic engineering—both were obtained by chemical mutagenesis. These mutants demonstrate the potential for large and unanticipated compositional changes as a result of genetic modifications by mutagenesis, a method other than genetic engineering.

In a separate set of experiments, different wild-type ecotypes (Col0 and Col24 [analogous to different plant varieties]) were studied to determine whether they could be distinguished by their metabolite profiles (Taylor et al., 2002). These two wild-type *Arabidopsis* are fully cross-fertile, and progeny from crosses between the two lines were also analyzed. Four hundred forty-three compounds were identified, but only 92 of these had structures unambiguously determined. Interestingly, the compounds showing the most variation were those whose structures and identities were known. The unknown compounds had on average tenfold less variation than known compounds. Col0 was lacking 27 peaks that were present in Col24, while Col24 was lacking 14 peaks that were present in Col0. Bioinformatics approaches to data analysis were able to differentiate the two wild-type ecotypes with relatively high precision.

A final example that highlights the biological variation that exists between and even within a plant is a metabolite profile study of pumpkin phloem exudates (Fiehn, 2003). Phloem is a specialized cell type in plants that transports a variety of nutrients (e.g., carbon fixed by photosynthesis in leaves) from source tissues (leaves) to sink tissues (root and fruit). As with prior metabolite profiling work, approximately 400 compounds (the majority unknown) could be identified. A surprising result of this work was that each leaf on an individual plant had a distinct overall metabolite profile that could be distinguished from the others by bioinformatics analyses.

The differences between identical aged leaves (e.g., leaf 2) of different plants grown under as identical conditions as possible also differed significantly. Approximately 30 to 40 percent of the metabolites of a leaf were significantly different from the overall average leaf profile. These data highlight the resolution that can be obtained with metabolite profiling, but they also raise issues of the natural biological variation that exists when tissues are considered for sampling. There is thus a need for studies that quantify the natural variation of the chemical composition of varieties used for nutrition before it is possible to establish if changes in modified varieties are within the normal range of variation.

LC-MS Analysis

LC is an alternative to GC for separating compounds for analysis. A major advantage to LC is that compounds do not have to be derivatized and can be resolved intact, although derivatization may still be needed to resolve particular compounds. LC separation can be coupled to a variety of detectors, including spectrometers. One example is the coupling of LC to a photodiode array detector that evaluates the ultraviolet/visible absorption spectra of a compound. This methodology has been used to profile isoprenoids (carotenoids, plastoquinones, and tocopherols) in tomato fruit (Fraser et al., 2000). However, photodiode array detectors are limited in the breadth of compounds detected, as many metabolites of interest do not absorb well in the visible or near ultraviolet wavelengths.

MS coupled with LC separation has many advantages over LC photodiode array approaches as many more classes of plant metabolic compounds (e.g., isoprenoids, alkaloids, flavanoids, and saponins) can be separated and detected with MS (Huhman and Sumner, 2002; Tolstikov and Fiehn, 2002). However, no single LC-MS procedure allows separation and determination of all classes of compounds, and the technique suffers from what are termed matrix effects, in which the presence of one compound in the spectrometer affects the ability to detect and quantify another compound. Unlike GC-MS, LC-MS methodology is still in its infancy. Standardized protocols and an understanding of factors affecting technical variation and reproducibility are still being developed by the scientific community.

NMR Spectroscopy

A third potential approach to profiling metabolites in an extract is NMR spectroscopy. NMR, in both proton and carbon-13 modes, can provide fingerprints with good chemical specificity for compounds that are present in relatively high abundance in a tissue or extract, but it has more limited utility for lower-abundance compounds. Unlike GC- and LC-based methods, NMR can be performed on whole tissues and, as such, need not be destructive. NMR can also be performed on extracts or fractionated extracts and interfaced with LC methodologies. Unlike LC-MS and GC-MS analyses, NMR can be performed relatively rapidly and with moderate to high throughput.

Several studies have shown the potential of NMR as a screening tool. Because plant constituents differ widely in their solubility properties, meaningful NMR analysis cannot be performed on a single plant extract. Extractions with multiple buffers or solvent mixtures to obtain groups of compounds of different polarity are required to broadly analyze plant components by NMR. In addition, due to biological variation, independent extracts of several plants need to be individually obtained and compiled to derive an average spectrum for a genotype.

Noteborn and colleagues (2000) examined proton NMR spectra of a genetically transformed tomato fruit and an appropriate control. In that study, extracts were prepared and analyzed in five fractions obtained by extraction with solvent mixtures of differing polarity. It was concluded that 27 to 30 percent of the detected compounds varied in concentration between transgenic and nontransgenic lines. However, most of these compounds were not identified. Although this study showed the potential of NMR for identifying compositional differences, the role of NMR in food safety assessment has not been demonstrated. A particular need is the standardization and optimization of extraction protocols and other aspects of sample handling (Defernez and Colquhoun, 2003; Kuiper et al., 2003).

Carbon-13 and proton NMR techniques also are powerful tools in investigating metabolic pathways and their inherent fluxes (Ratcliffe and Shachar-Hill, 2001; Roberts, 2000). In this context, an appropriate metabolic precursor that is labeled with one or more carbon-13 or deuterium atoms is introduced into the organism where it can undergo further metabolism. Analysis of various extracts can identify the intermediates and products of metabolism and, if done sequentially over time, rates of reactions can be determined. While this does not provide information regarding composition, which is the primary analytical goal in a food safety assessment, such techniques do provide a means of assessing metabolic changes that might be introduced by genetic engineering, mutagenesis, or conventional breeding.

BIOINFORMATIC ISSUES IN PROFILING ANALYSIS

Metabolic fingerprinting is carried out by any method that can provide a pattern that is unique to a certain sample. Such methods do not necessarily attempt to identify any compounds (or proteins or DNA). In addition to the methods involving a chromatographic separation (e.g., GC-MS and LC-MS), methods that are often used for fingerprinting are direct spectral techniques. These include: NMR (Holmes et al., 2000; Raamsdonk et al., 2001); Fourier transform infrared spectroscopy (Johnson et al., 2003); and direct infusion mass spectrometry, where samples do not pass through a preliminary separation such as GC or LC (Goodacre et al., 2002).

In all of these cases the spectra are composed of peaks that may not necessarily correspond to chemical species, as they may form from their interactions (e.g., the phenomena of ion suppression in direct-infusion MS). Nevertheless, the spectra are characteristic of the sample. Such metabolic fingerprinting is mostly useful for purposes of classifying samples, such as determining whether two samples are chemically different. Classification is made by applying pattern recognition algorithms to a set of training data, for which one must know the class membership of all samples.

Many algorithms exist that are suitable for this task, such as discriminant function analysis, artificial neural networks, hidden-Markov models, decision trees, genetic programming, and other statistical and machine-learning methods. Essentially, these methods are first calibrated with the training data set so as to optimally distinguish, through the use of a combination of the spectral variables, samples that belong to different classes.

After calibration, such algorithms can be used to classify whether the unknowns are similar to or different from the training set. This particular approach of fingerprinting is more robust to artifacts arising from technical variation (e.g., in sample preparation) than the targeted approaches that are based on identification of particular chemicals in the sample. While fingerprinting may be well-suited to distinguishing samples containing particular sources of food by detecting compositional differences (e.g., genetic engineering versus isogenic control), it is inappropriate for identifying unintended effects of genetic modification because there is no attempt to determine the specific chemical nature of the compositional change identified, only that one sample is different from another.

One particular approach that has great potential in identifying possible unintended effects of genetic modification is based on comparison with baseline metabolic profiles. The metabolic profile (obtained by GC-MS or LC-MS) of a GE organism (GEO) would be compared with the profiles of corresponding wild-type organisms, and those peaks that differ significantly would be identified (in particular, the appearance of new peaks, but also the absence or reduction of existing peaks).

Those components that appear in significantly different amounts in the modified organism as compared with the baseline would then need to be identified chemically, bearing in mind that even in plants that have not been genetically engineered, only about 20 percent of the compounds detected by the method can be associated with a known structure. The next step would be to assess whether there is a possibility of toxicity or other negative biological effect, which is also a technically daunting task (see Chapter 6).

With the baseline metabolic profile approach, an interval profile that is representative of the wild-type organism must be created. To do so requires measuring a large number of samples to establish limits of natural variation for each component (peak) of the profile. This may include profiling several different strains, varieties, or breeds when these are common in the diet. An important detail of this approach is that the statistical significance of the differences needs to be established, which also requires several samples of the modified organism and application of appropriate statistical methods, because it is important to show that the difference between the profile of the modified organism is larger than the normal range of variation of the wild-type samples measured. The number of biological replicates for wild type and the GM crop should be large (at least several dozen) to provide ample data sets to describe the biological variation and thus provide sufficient statistical power. A database management system should also be constructed specifically for this purpose, and it needs to capture as much metadata (data about the experimental protocol) as possible.

The documentation of detailed procedures is the only way to demonstrate that the comparisons were made with a minimal chance of introducing systematic errors that may have generated false levels of similarity or differences between wild-type and modified sample populations. Such a database of interval profiles for various food sources would then be a primary way to assess differences in composition and would also be an important resource to establish the nutritional value of different food sources (literally the biochemical difference between apples and oranges).

Implications for Predicting Unintended Health Effects

Increased analytical capability does not equate to an enhanced ability to predict health outcomes for several reasons. For even the best-studied plant systems (e.g., *Arabidopsis*), only a fraction of the compounds present can be resolved by a given method (e.g., LC-MS or GC-MS). Of those that are resolved, only a fraction (20-30 percent) can be identified with certainty as to their structure. The vast majority of chemical peaks detected by these methods remain classified as unknown compounds.

It is very difficult to interpret or predict the effects on human health of changes in the composition of a single food item and, especially, the health effects of changes in a single food item present in the total diet. This is true even for compounds for which large amounts of nutritional data are already known (e.g., specific amino acids or fatty acids; see Chapter 6). Problems in assessing the significance to human health of compositional changes in individual food or in the total diet are further amplified by the biological variation between samples, differences in analytical protocols and results between laboratories, and changes in composition that inevitably occur over time. Although advanced technologies are promising, limited knowledge of their role in mammalian systems, along with an inability to identify or functionally characterize differences, prohibit them from serving as a basis of safety assessment at this time. Their most useful present application may be the detection and quantification of known toxic compounds.

Interpretive Limitations of Metabolite
Separation and Analysis Techniques

The past decade has seen incredible advances in the technologies associated with analytical separation of plant metabolites. The combined advances in applying GC-MS and, more recently, LC-MS and NMR to characterize plant metabolism have increased by a factor of 10 to 100 the number of compounds that can be resolved with modern instrumentation. Even though some analyses may yield up to 1,000 different chemical species, this is still a fraction of the total number of compounds present in a plant extract. The parallel combination of GC-MS and LC-MS holds promise for allowing the future analysis of a much larger percentage of metabolites in a plant extract. Both methodologies are highly demanding from an analytical perspective and are prone to the generation of artifacts unless conditions of extraction, preparation, and analysis are rigidly standardized and enforced on a global scale.

A serious limitation in these analyses, even if the numbers of compounds that can be resolved and identified increases from 1,000 to 5,000, is that a large number of compounds will remain to be identified by any of the procedures described above. For GC-MS analyses, 50 to 80 percent of potentially unidentified compounds are still unknown. A major international effort is required to chemically and structurally characterize the compounds in plants and build plant compound-specific spectral libraries and reference databases to address this issue. The development of internationally recognized and followed standards for extraction, derivatization, chromatographic separation, and detection is needed to allow data from different laboratories to be compared across space and time. The validation of separation and quantification methodologies using agreed-upon standards is also needed to ensure reproducibility and comparison.

Pattern Recognition Methods for Evaluation of Compositional Equivalence

The profiling methods described above are useful analytical tools for unique compounds. However, they have limitations when applied to complex mixtures, such as food. Food, particularly plant-derived food, is a mixture of thousands of different compounds. Many of these compounds will coelute in the analysis, even though they are different compounds. As the complexity of the mixture increases, there is a greater probability that unique compounds will not be identified by currently available profiling techniques. Thus additional analytical tools must be identified and applied to screen complex mixtures for unique compounds that may initiate an adverse health effect when consumed.

An additional consideration is the biological relevance of a new compound. Unless each individual compound in a new modified food is tested for adverse effects, the biologic relevance will not be identified (see Chapter 5). Determining biologic relevance requires analysis beyond profiling individual compounds. Generally animal models are used to detect adverse effects from new compounds. An adverse response to a new compound may be seen when it is tested as a pure compound, but this is not always the case when it is tested as a component in a mixture, such as a food. An example is the introduction of a nutrient, such as iron, into a food that contains chelating agents, such as certain polysaccharides. When tested as a pure compound, iron will have greater biologic activity than when tested in a mixture that contains chelating agents that will bind and thereby decrease its biologic activity.

An additional consideration is the level of a compound that is introduced into a test animal to detect adverse effects. Again, when the compound is pure, higher levels can be tested in in vivo animal systems than can be introduced as a food component. Thus adverse effects may be detected with high levels of a pure compound that would not be seen at low levels and in a complex mixture.

PROFILING METHODS FOR ANALYSIS OF INORGANIC ELEMENTS OF NUTRITIONAL AND TOXICOLOGICAL IMPORTANCE

The major focus of discussion has been the analysis of organic compounds; however, the analysis of trace elements poses equally important but distinctly different challenges. Unlike the vast array of organic compounds present in foods, the inorganic constituents of foods constitute a much smaller array of analytes to be measured. As discussed in Chapter 6, many mineral elements are nutritionally essential but have toxic potential at only slightly higher levels of intake, and interactions can occur among the nutritionally essential minerals. In addition, changes in plant genetics, especially modifications intended to alter mineral concentration (Clemens et al., 2002; Holm et al., 2002) have the potential to alter the content of multiple trace elements. Thus, it is essential that the focus of mineral analysis not be overly narrow.

The traditional targeted analytical approach for the measurement of inorganic elements is generally based on sensitive and specific methods such as atomic absorption or atomic emission spectrophotometry. Although targeted analyses are fully adequate for the determination of individual elements of nutritional and toxicological interest, analytical approaches that allow a determination of multielement profiles have conceptual and practical advantages for monitoring the composition of GM food products.

Nontargeted mineral analysis is performed typically using either inductively coupled plasma-optical emission spectrometry or inductively coupled plasma-mass spectrometry, with thermal ionization mass spectrometry as an alternative in some applications. These techniques, with proper attention to sample preparation and method calibration, are capable of providing quantitative data in profiling a wide range of elements, including aluminum, iron, potassium, magnesium, sodium, lead, zinc, arsenic, cadmium, calcium, molybdenum, cobalt, chromium, copper, mercury, manganese, nickel, tin, selenium, strontium, and vanadium in foods and many other types of samples (Almeida et al., 2002; Brescia et al., 2003; Cariati et al., 2003; Frachler et al., 1998; Losso et al., 2003; Wang et al., 2000). These applications provide quantitative data on the total content of the elements in a sample, while coupling multielement analysis with a preliminary separation of proteins yields information about the metal content of specific metal-binding proteins such as metallothioneins (Goenaga Infante et al., 2003).

As discussed in the context of organic constituents, a critical aspect of multielement analysis is the interpretation of data. Subtle differences in patterns of inorganic elements can be indicative of the geographic origin of agricultural commodities (Brescia et al., 2003) but may have little or no nutritional or toxicological significance. As with other aspects of the profiling of food constituents, the interpretation of data from multielement analysis is the critical issue in evaluating differences due to genetic modification in the context of compositional effects of geographic location, climate, and agronomic variables.

GENOMICS

It has been proposed that differential gene expression be used as a method to determine the substantial equivalence of genetically modified organisms (GMOs), including between genetically engineered organisms (GEOs) and other GMOs (GAO, 2002; Kuiper et al., 2001, 2003; van Hal et al., 2000). Genomic technologies can measure the level of thousands of transcripts simultaneously, thereby providing a molecular phenotype that can be used to compare transcript expression between the immediate progenitor and the GM species. Differential gene expression can be measured using open and closed technologies (Green et al., 2001).

Open technologies, such as serial analysis of gene expression, do not require prior sequence knowledge of the organism, can survey all transcripts of the organism in a given tissue under study, can capture transcript sequence information (e.g., splice variations or small nucleotide polymorphisms), and are quantitative. However, open systems require extensive DNA sequencing to achieve a critical mass of data to adequately profile gene expression of the organism and are not likely to be cost-effective for routine screening. Serial analysis of gene expression has been used to identify differentially expressed genes in rice (Matsumura et al., 1999).

In contrast, closed systems, such as GeneChips or cDNA/oligonucleotide microarrays, require a priori sequence information for each gene that is to be monitored (van Hal et al., 2000). Microarrays only measure the expression of genes represented on the array and, in general, do not adequately account for differences resulting from naturally occurring differences in a gene sequence between organisms (splice variants). In addition, closed systems are specific to the organism and, to some extent, the strain. Analyses of data for both open and closed systems are still emerging, with different approaches significantly affecting outcome, interpretation, and the conclusions drawn (Quackenbush, 2002).

There are a number of challenges that must be addressed before incorporating microarray technology into the safety assessment of GM food. One major issue that limits the utility of differential gene expression technology to assess substantial equivalence is the lack of data regarding the expression level of genes in an organism under various growth conditions and developmental stages, as well as in cells and tissues (GAO, 2002). Ranges of these expression levels must be defined. Furthermore there is a questionable correlation between the level of a transcript and the gene product. Therefore, differences in gene expression between the progenitor and the GMO may not be reflected in differences in the level of expressed protein.

PROTEOMICS

The term proteomics ideally refers to the analysis of the complete complement of proteins of an organism. In practical terms it is still impossible to detect all proteins of an organism but, at least for model organisms, a large proportion of the predicted proteins can be detected. Proteins are extremely important biochemical components as they are very abundant in all biological material, and they are the molecular machines that function in cells. They are made up of linear chains of any of 21 individual amino acid units (plus a few more very rare amino acids) that occur in varying quantities, patterns, and fold in characteristic, but diverse, secondary and tertiary structures. Proteins are also the major component of the human immune system and are able to recognize other proteins by binding.

In general, proteins are broken down into small peptides and amino acids in the digestive tract and so their amino acid composition is important in human and animal nutrition. However, some proteins are very stable and resist digestion, while others are detected by the immune system at extremely low levels and cause severe allergic reactions in a proportion of humans and animals. Analyzing the constituent proteins of plants and animals for human consumption is an important component of assessing the consequences of genetic or other modifications. An additional factor is that assessing changes in the composition of particular proteins may reveal changes in chemical compounds that are not detectable or identifiable with the techniques of metabolomics.

Detection and Identification of Proteins

Proteomic analysis differs in both the techniques used and the analytical intent from the targeted analyses (e.g., HPLC, enzyme-linked immunosorbant assay, and immunoblot techniques) used to detect and quantify individual proteins or groups of proteins. Protein identification is almost exclusively done by MS methods, while quantification is done through 2-dimensional gel electrophoresis or tandem-LC. In all cases, the protein separation precedes the MS identification step.

As recently reviewed by Regnier and colleagues (2002), a growing application in protein analysis is termed comparative proteomics, in which a reference sample is compared with a sample derived from an altered state. Comparative proteomics has great potential in the evaluation of the compositional effects of genetic changes in food and other biological material. However, at present the technology has not been sufficiently well developed for use in the routine assessment of GE food. As with other forms of profiling analysis, a major problem is the issue of data processing and bioinformatics (Regnier et al., 2002). For example, how effectively and reliably can any proteomic technique identify and quantify changes in a protein of potential toxicological interest among thousands of other proteins in the sample?

Comparative proteomic analysis can be performed by either of two basic approaches (Regnier et al., 2002). The traditional approach is the separation of intact proteins in a sample. Two-dimensional polyacrylamide gel electrophoresis is the original and probably remains the most common application of such a mass separation approach.

Separation of proteins in 2-dimensional gel electrophoresis is a technique that has been practiced in laboratories for nearly 30 years, since its development in 1975 (O'Farrell, 1975). Proteins are separated first by their acidic properties in one direction and then by their size in the orthogonal direction in the polyacrylamide gel. The gel is then stained with a dye that reveals where protein spots are located; the protein abundance is calculated based on the size and intensity of the spot. Gels usually resolve several hundred to a few thousand different protein spots from a biological matrix. After visualizing the protein spots, they are identified by MS using one of two methods. (A few other methods have been proposed, but their development is still at a very early stage and their use is not realistic in production environments.)

The initial interpretation generally involves a comparison of patterns among reference and test sample and, thus, is similar in principle to other profiling methods in which the pattern is analyzed without knowledge of the identity of most of the components. The identity of unknown proteins (i.e., spots on a 2-D gel electrophoresis) can frequently be determined by further analysis of a partial amino acid sequence obtained by chemical or MS analysis.

MS analysis of proteins (e.g., from 2-D gel electrophoresis) is based on the assumption that an adequate database of amino acid and nucleotide sequences exists for the plant being analyzed or that sufficient homology exists with more fully characterized species. This issue is a major limitation in the analysis of many plant proteins at this time.

When sequence information does lead to a tentative identification, the identity of the unknown can be supported by additional information (mass and iso-electric point) derived from protein separation by electrophoretic mobility. This technique is, at best, semiquantitative. In addition, the identification of poten-tially important differences among reference and test samples is complicated by the extreme complexity of the array of expressed proteins, which is further com-plicated by post-translational modification. Although this traditional approach to proteomic analysis has been used to characterize many proteins in plant systems, there has been little or no application of protein patterns in comparing plants or animals that have been altered by genetic engineering or other variables that would affect protein expression.

When the genome of the organism under analysis has been fully sequenced as, for example, for baker's yeast and rice, the method of peptide mapping can be used to identify proteins. This method consists of breaking the proteins in each gel spot with a single endoprotease enzyme that cuts the sequence of amino acids at well-known positions, resulting in a mixture of smaller peptides. The peptide mixture is then injected into a mass spectrometer; usually a matrix-assisted laser desorption/ionisation-time of flight mass spectrometry instrument, which can ac-curately measure the mass of the peptides (down to fractions of atomic units), and the amino acid composition of the peptide is then calculated. The full genome of that organism would have been previously scanned to list all protein sequences it encoded, and these would be used to predict the products of digestion by the endoprotease used in the assay and to accurately calculate their masses.

Next, the proteins are identified by matching the accurate peptide masses measured with the theoretical masses of the peptides obtained from the genome sequence. This method is routine in a growing number of research laboratories and core facilities. When the full genome sequence of the organism has not been determined, which is true for the majority of species of nutritional importance, a different strategy for protein identification must be used. This requires using mass spectrometers that interface with LC (quadrupole-time of flight or, essentially, ion traps) and that are able to perform tandem MS so that the peptides can be broken one amino acid at a time and the accurate mass of the resulting peptides can be measured. A partial amino acid sequence is thus obtained for each of the constituent peptides of the original protein, and identification is carried out by finding other known proteins whose sequence is similar to the partial sequences obtained from the mass spectrometer. Both methods are absolutely dependent on bioinformatic methods and on complete genome-sequencing efforts.

The alternative approach in proteomic analysis is termed peptide mapping. In this approach the proteins in the sample are subjected to partial hydrolysis prior to any separation, and this is followed by GC-MS or LC-MS analysis (Regnier et al., 2002). The resulting pattern is an extremely complex mixture of peptides that may reflect compositional differences among samples. Determining the identity of the differing proteins in this method requires that a database of predicted peptides be created for each organism. This is a one-time effort and is not a limiting factor in using this technique as all equipment vendor software comes equipped with such databases. However, a limitation does exist in the current state of the art in annotation of gene function. While all of the proteins of rice may be able to be identified, more than half of these are of unknown function. Similar ratios are typical from other plants and animals of nutritional interest. The method of identification by partial protein sequencing requires a more involved bioinformatic approach. In this case the identification relies on the similarity of sequence between the protein of interest and other known proteins of any origin. Thus the reference sequence database should include all known proteins and be kept up to date.

Usually the databases used in proteomic analysis are the Swiss-Prot (a database with high-quality annotation) and TrEMBL (a computer-annotated supplement to Swiss-Prot) combination (Boeckmann et al., 2003; http://www.ebi.ac.uk/swissprot/) or the nonredundant version of GenBank (translated to protein sequence code). One improvement to this approach is to use a focused database of partial-known sequences that derives from expressed sequence tag (EST) projects. EST projects exist for many species of nutritional interest and, in several cases these are large bodies of expressed sequences (e.g., soybean or corn). The idea then is to attempt to match the partial peptide sequences to the partial mRNA sequences. Matches allow the relation of the protein to the one particular mRNA fragment, but identification is still ultimately done through a match to all known protein sequences because the EST is also annotated in the same way. Proteomics is indeed most effective with fully sequenced genomes.

An issue that is problematic in many current applications of proteomics relates to the inefficiencies of the 2-dimensional gel separation. One problem is that a number of proteins do not migrate well in these gels, membrane proteins being the most abundant of this class. A second issue is related to the limit of detection of the staining process that reveals where proteins are located in the gel. The commonly used Coomassie-blue dye has a rather high limit of detection, resulting in many protein spots in the gel never being revealed in the analysis.

Alternatives to this dye are silver staining and fluorescently labeled stains. While silver is problematic because of interference with the MS processes, the use of fluorescence dyes has lowered the limit of detection. However, the dynamic range of these dyes is also not as good as needed. As a result, it is still not possible to accurately measure the very abundant proteins or the very low abundant proteins. This is perhaps the greatest obstacle to using proteomics to monitor unintended effects of modification because strongly allergen proteins are often present in very low concentrations.

What Should be Analyzed?

Proteomics, like metabolomics, can be used either to profile proteins, resulting in lists of proteins present in the analyte, or to fingerprint, where only a characteristic signature of the biological matrix is obtained. The latter can be effective in terms of comparing modified organisms with the wild types, but it does not necessarily identify the sources of difference. This may be a faster first-screening process, which can be followed by more detailed profiling when changes exceed a defined threshold. As stated previously, immunochemical assays for specific proteins of interest (e.g., known allergens) can be incorporated into targeted analysis independent of a proteomic analysis.

INFORMATION OBTAINED FROM
NEW ANALYTICAL TECHNIQUES

With the increased sensitivity and resolution of technology during the past decade, an analyst now has the ability to detect and quantify tens of thousands of possible changes in biological molecules (e.g., DNA, RNA, protein, and metabolites) in a given system. Complete genome sequences have been obtained for many organisms, and this allows scientists to identify nearly all the genes (protein encoding and otherwise) in an organism. However, the majority of the proteins encoded by the genes in an organism are novel to biology and their functions remain unknown. An understanding of the physiological, developmental, and biochemical roles and interrelationships of selected genes for the growth and development of an organism are known for only a small fraction of the genes in a given genome.

Complete genome sequences and other technological advances have driven the development of techniques that make it possible to simultaneously analyze the expression of many thousands of genes in an organism in a particular tissue, such as the time of development and growth. However, the ability to identify which changes in gene expression are biologically significant and to place this global view of gene expression into a biological context is quite limited, even for the best-studied organisms.

The situation for the global analysis of protein expression levels and protein modifications (proteomics) is even less advanced, as the protein complement of an organism is more complex than DNA because proteins can be post-translationally modified and exist in several different forms, not all of which may have the same function in a cell. Finally, new techniques have allowed characterization of changes in the levels and types of a wide variety of biochemical compounds. Again, in this case no technique is available that can provide a complete characterization of all molecules in a cell or tissue. Even if this were possible, the vast majority of metabolites observed have not been identified chemically, and the biological significance for the organism in which the compounds are produced or for other organisms that ingest these compounds as part of their diet remains unknown.

Thus while new global technologies for profiling gene expression, proteins, and metabolites have increased the breadth and resolution of analyses that are possible in biological systems and now allow scientists to generate vast expression, protein, and metabolite datasets for a single tissue, our ability to relate these vast datasets in a predictive food safety context remains limited. Analytical capabilities have increased substantially during the past decade, but these have not been accompanied by parallel increases in the ability to understand the biological consequences of individual compounds or complex mixtures of compounds or by the ability to predict adverse health effects from exposure to new compounds in food. The chemical identity and biological relevance of a large percentage of new compounds that may be identified by the methods described in this chapter are unknown.

DNA sequencing can, however, provide the full complement of the genetic information encoded in an organism. For transgenic organisms, DNA sequencing allows the precise location of the inserted transgene in the genome and the context of the inserted gene to be determined. Thus it can readily be determined if the transgene inserted has disrupted a gene encoded in the organism.

DISCUSSION

Regardless of the analytical methods used and the quality and depth of compositional data obtained, data interpretation remains the critical issue for evaluating the significance of unintended compositional changes. Questions that bear consideration include the following: How is analytical data for a new food, whether genetically engineered or produced by conventional methods, interpreted? Against what references should the new food be compared? Should the same analytical path be used for GE and non-GE food? In Chapter 7, an analytical framework is proposed for addressing these and related issues.

REFERENCES

Almeida CM, Vasconcelos MT, Barbaste M, Medina B. 2002. ICP-MS multi-element analysis of wine samples—A comparative study of the methodologies used in two laboratories. *Anal Bioanal Chem* 374:314–322.

Ashihara H, Crozier A. 2001. Caffeine: A well known but little mentioned compound in plant science. *Trends Plant Sci* 6:407–413.

Bakan B, Melcion D, Richard-Molard D, Cahagnier B. 2002. Fungal growth and fusarium mycotoxin content in isogenic traditional maize and genetically modified maize grown in France and Spain. *J Agric Food Chem* 50:728–731.

Boeckmann B, Bairoch A, Apweiler R, Blatter MC, Estreicher A, Gasteiger E, Martin MJ, Michoud K, O'Donovan C, Phan I, Pilbout S, Schneider M. 2003. The SWISS-PROT protein knowledgebase and its supplement TrEMBL in 2003. *Nucleic Acids Res* 31:365–370.

Brescia MA, Kosir IJ, Caldarola V, Kidric J, Sacco A. 2003. Chemometric classification of Apulian and Slovenian wines using 1H NMR and ICP-OES together with HPICE data. *J Agric Food Chem* 51:21–26.

Cariati F, Fermo P, Gilardoni S. 2003. Optimization of an urban particulate matter multi-element analysis method by inductively coupled plasma-atomic emission spectrometry (ICP-AES). *Ann Chim* 93:539–550.

Clemens S, Palmgren MG, Krämer U. 2002. A long way ahead: Understanding and engineering plant metal accumulation. *Trends Plant Sci* 7:1360–1385.

Clements MJ, Campbell KW, Maragos CM, Pilcher C, Headrick JM, Pataky JK, White DG. 2003. Influence of Cry1Ab protein and hybrid genotype on fumonisin contamination and fusarium ear rot of corn. *Crop Sci* 43:1283–1293.

Defernez M, Colquhoun IJ. 2003. Factors affecting the robustness of metabolite fingerprinting using 1H NMR spectra. *Phytochemistry* 62:1009–1017.

Dowd PF. 2001. Biotic and abiotic factors limiting efficacy of *Bt* corn in indirectly reducing mycotoxin levels in commercial fields. *J Econ Entomol* 94:1067–1074.

Duvick J. 2001. Prospects for reducing fumonisin contamination of maize through genetic modification. *Environ Health Perspect* 109S:337–342.

FAO/WHO (Food and Health Organization of the United Nations/World Health Organization). 2003. Report of the Fourth Session of the Codex *Ad Hoc* Intergovernmental Task Force on Foods Derived from Biotechnology, Yokohama, Japan. Online. Available at ftp://ftp.fao.org/docrep/fao/meeting/006/y9220e.pdf. Accessed September 21, 2003.

Fiehn O. 2002. Metabolomics: The link between genotypes and phenotypes. *Plant Mol Biol* 48:155–171.

Fiehn O. 2003. Metabolic networks of *Cucurbita maxima* phloem. *Phytochemistry* 62:875–886.

Fiehn O, Kopka J, Dörmann P, Altman T, Trethewey RN, Willmitzer L. 2000a. Metabolite profiling for plant functional genomics. *Nat Biotechnol* 18:1157–1161.

Fiehn O, Kopka J, Trethewey RN, Willmitzer L. 2000b. Identification of uncommon plant metabolites based on calculation of elemental compositions using gas chromatography and quadrupole mass spectrometry. *Anal Chem* 72:3573–3580.

Frachler M, Rossipal E, Irgolic KJ. 1998. Trace elements in formulas based on cow and soy milk and in Austrian cow milk determined by inductively coupled plasma mass spectrometry. *Biol Trace Elem Res* 65:53–74.

Fraser PD, Pinto ES, Holloway DE, Bramley PM. 2000. Application of high-performance liquid chromatography with photodiode array detection to the metabolic profiling of plant isoprenoids. *Plant J* 24:551–558.

GAO (General Accounting Office). 2002. *Genetically Modified Foods: Experts View Regimen of Safety Tests as Adequate, but FDA's Evaluation Process Could Be Enhanced.* GAO-02-566. Washington, DC: GAO.

German JB, Roberts MA, Watkins SM. 2003. Genomics and metabolomics as markers for the interaction of diet and health: Lessons from lipids. *J Nutr* 133:2078S–2083S.

Goenaga Infante H, Van Campenhout K, Schaumlöffel D, Blust R, Adams FC. 2003. Multi-element speciation of metalloproteins in fish tissue using size-exclusion chromatography coupled on-line with ICP-isotope dilution-time-of-flight-mass spectrometry. *Analyst* 128:651-657.

Goodacre R, Vaidyanathan S, Bianchi G, Kell DB. 2002. Metabolic profiling using direct infusion electrospray ionisation mass spectrometry for the characterisation of olive oils. *Analyst* 127:1457–1462.

Green CD, Simons JF, Taillon BE, Lewin DA. 2001. Open systems: Panoramic views of gene expression. *J Immunol Methods* 250:67–79.

Gregory JF. 1998. Nutritional properties and significance of vitamin glycosides. In: McCormick DB, ed. *Annu Rev Nutr.* Palo Alto, CA: Annual Reviews. Pp. 277–296.

Holm PB, Kristiansen KN, Pedersen HB. 2002. Transgenic approaches in commonly consumed cereals to improve iron and zinc content and bioavailability. *J Nutr* 132:514S–516S.

Holmes E, Nicholls AW, Lindon JC, Connor SC, Connelly JC, Haselden JN, Damment SJ, Spraul M, Neidig P, Nicholson JK. 2000. Chemometric models for toxicity classification based on NMR spectra of biofluids. *Chem Res Toxicol* 13:471–478.

Huhman DV, Sumner LW. 2002. Metabolic profiling of saponins in *Medicago sativa* and *Medicago truncatula* using HPLC coupled to an electrospray ion-trap mass spectrometer. *Phytochemistry* 59:347–360.

IOM (Institute of Medicine). 2000. Vitamin E. In: *Dietary Reference Intakes for Vitamin C, Vitamin E, Selenium, and Carotenoids.* Washington, DC: National Academy Press. Pp. 186–283.

Johnson HE, Broadhurst D, Goodacre R, Smith AR. 2003. Metabolic fingerprinting of salt-stressed tomatoes. *Phytochemistry* 62:919–928.

Kuiper HA, Kleter GA, Noteborn HP, Kok EJ. 2001. Assessment of the food safety issues related to genetically modified foods. *Plant J* 27:503–528.

Kuiper HA, Kok EJ, Engel KH. 2003. Exploitation of molecular profiling techniques for GM food safety assessment. *Curr Opin Biotechnol* 14:238–243.

Lee SJ, Ahn JK, Kim SH, Kim JT, Han SJ, Jung MY, Chung IM. 2003. Variation in isoflavone of soybean cultivars with location and storage duration. *J Agric Food Chem* 51:3382–3389.

Liu CJ, Blount JW, Steele CL, Dixon RA. 2002. Bottlenecks for metabolic engineering of isoflavone glycoconjugates in *Arabidopsis. Proc Natl Acad Sci U S A* 99:14578–14583.

Losso JN, Munene CN, Moody MW. 2003. Inductively coupled plasma optical emission spectrometric determination of minerals in catfish frame. *Nahrung* 47:309–311.

Matsumura H, Nirasawa S, Terauchi R. 1999. Technical advance: Transcript profiling in rice (*Oryza sativa L.*) seedlings using serial analysis of gene expression (SAGE). *Plant J* 20:719–726.

Noteborn HPJM, Lommen A, van der Jagt RC, Weseman JM. 2000. Chemical fingerprinting for the evaluation of unintended secondary metabolic changes in transgenic food crops. *J Biotechnol* 77:103–114.

O'Farrell PH. 1975. High resolution two-dimensional electrophoresis of proteins. *J Biol Chem* 250:4007–4021.

Quackenbush J. 2002. Microarray data normalization and transformation. *Nat Genet* 32S:496–501.

Raamsdonk LM, Teusink B, Broadhurst D, Zhang N, Hayes A, Walsh MC, Berden JA, Brindle KM, Kell DB, Rowland JJ, Westerhoff HV, van Dam K, Oliver SG. 2001. A functional genomics strategy that uses metabolome data to reveal the phenotype of silent mutations. *Nat Biotechnol* 19:45–50.

Ratcliffe RG, Shachar-Hill Y. 2001. Probing plant metabolism with NMR. *Annu Rev Plant Physiol Plant Mol Biol* 52:499–526.

Regnier FE, Riggs L, Zhang R, Xiong L, Liu P, Chakraborty A, Seeley E, Sioma C, Thompson RA. 2002. Comparative proteomics based on stable isotope labeling and affinity selection. *J Mass Spectrom* 37:133–145.

Roberts JKM. 2000. NMR adventures in the metabolic labyrinth within plants. *Trends Plant Sci* 5:30–34.

Roessner U, Wagner C, Kopka J, Trethewey RN, Willmitzer L. 2000. Technical advance: Simultaneous analysis of metabolites in potato tuber by gas chromatography-mass spectrometry. *Plant J* 23:131–142.

Roessner U, Luedemann A, Brust D, Fiehn O, Linke T, Willmitzer L, Fernie A. 2001. Metabolic profiling allows comprehensive phenotyping of genetically or environmentally modified plant systems. *Plant Cell* 13:11–29.

Saxena D, Stotzky G. 2001. *Bt* corn has a higher lignin content than non-*Bt* corn. *Am J Bot* 88:1704–1706.

Shintani D, DellaPenna D. 1998. Elevating the vitamin E content of plants through metabolic engineering. *Science* 282:2098–2100.

Taylor J, King RD, Altmann T, Fiehn O. 2002. Application of metabolomics to plant genotype discrimination using statistics and machine learning. *Bioinformatics* 18:S241–S248.

Thelen JJ, Ohlrogge JB. 2002. Metabolic engineering of fatty acid biosynthesis in plants. *Metab Eng* 4:12–21.

Tolstikov VV, Fiehn O. 2002. Analysis of highly polar compounds of plant origin: Combination of hydrophilic interaction chromatography and electrospray ion trap mass spectrometry. *Anal Biochem* 301:298–307.

van Hal NL, Vorst O, van Houwelingen AM, Kok EJ, Peijnenburg A, Aharoni A, van Tunen AJ, Keijer J. 2000.The application of DNA microarrays in gene expression analysis. *J Biotechnol* 78:271–280.

Verpoorte R, Memelink J. 2002. Engineering secondary metabolite production in plants. *Curr Opin Biotechnol* 13:181–187.

Wall JS, Young MR, Carpenter KS. 1987. Transformation of niacin-containing compounds in corn during grain development: Relationship to niacin nutritional availability. *J Agric Food Chem* 35:752–758.

Wang T, Wu J, Hartman R, Jia X, Egan RS. 2000. A multi-element ICP-MS survey method as an alternative to the heavy metals limit test for pharmaceutical materials. *J Pharm Biomed Anal* 23:867–890.

Yan L, Kerr P. 2002. Genetically engineered crops: Their potential use for improvement of human nutrition. *Nutr Rev* 60:135–141.

5

Adverse Impacts of Food on Human Health

This chapter focuses on the range of health hazards, both documented (e.g., microbial) and perceived (e.g., due to the inadvertent mixture of various grains or to the consumption of deoxyribonucleic acid [DNA]), that can be associated with food, whether or not produced by biotechnology. It starts with an overview of food safety issues in general and then describes the context in which new genetically engineered (GE) foods are entering the market. This is followed by a description of the array of potential hazards that should be considered by efforts designed to anticipate or evaluate unintended adverse health effects. It is important to emphasize that the hazards presented below can occur with foods regardless of the method of production or processing and are not specific to the process of genetic engineering.

INTRODUCTION

Predicting and assessing potential adverse human health impacts arising from compositional changes in foods modified by a number of methods, including the genetic engineering of foods, are challenging. Adverse consequences could be narrow in occurrence or diverse and widespread and, because they are unintended, will be unexpected. Foods that could be modified in composition as a result of agricultural biotechnology, as defined in Chapter 1 and described in Chapters 2 and 3, are of interest because of the growing awareness that commonly consumed food constituents and complex mixtures can be beneficial or harmful to health.

Estimates based on population-based research indicate that approximately one-third of preventable morbidity and mortality is of dietary origin and/or a consequence of low levels of physical activity. In contrast to such long-term consequences, acute toxicities of dietary origin appear to pose a relatively small population health burden. Acute food toxicities may be very severe, but they generally affect much smaller numbers of people and can be associated rapidly with the food source, so that they usually can be controlled relatively easily.

FOOD SAFETY HAZARDS IN FOOD PRODUCTS

General Hazards from Foods

A variety of safety hazards are associated with foods produced by any method. These can be categorized from greatest to least hazardous by their probability to cause an adverse health effect as:

1. pathogenic microorganisms,
2. nutrient imbalances,
3. naturally occurring toxicants,
4. environmental and industrial chemicals, including pesticides,
5. food and feed additives,
6. food alterations associated with genetic modification.

This categorization was first proposed by Wodicka (1982).

Pathogens

Types of Pathogenic Microorganisms

Pathogenic microorganisms in food include: viruses, bacteria, toxin-producers, and parasites. Food-borne pathogens are often particularly risky for children, the elderly, and the immune-suppressed. There are millions of people stricken by food-borne illness every year in the United States and an estimated 76 million illnesses, 325,000 hospitalizations, and 5,000 deaths per year, mostly among the elderly and the very young (CDC, 2003).

In the United States, the Norovirus is the most commonly found cause of food-borne illness; other viruses (rotavirus and astrovirus), as well as parasites (*Giardia*) and bacteria (*Campylobacter*), play a major role. Three pathogens, *Salmonella, Listeria monocytogenes*, and *Toxoplasma gondii*, are responsible for 1,500 deaths each year; other pathogens that also contribute to morbidity and mortality due to food-borne pathogens include Norovirus, *Campylobacter,* and *Escherichia coli* O157:H7.

It is estimated that unknown pathogenic agents account for 81 percent of illnesses and hospitalizations and 64 percent of deaths due to food-borne illness (Mead et al., 1999). These numbers are far lower than in the past; in the United

States, measures such as drinking water disinfection, sewage treatment, milk sanitation and pasteurization, and shellfish monitoring have been largely successful. Newly emerging food safety hazards, however, are largely attributed to foodborne zoonoses that do not necessarily cause illness in animals and are therefore difficult to detect. Additionally, new vehicles have been identified, for example, *Salmonella enteritidis* found inside eggs (and not just on shells) and bacterial contaminants in juices, fruits, and vegetables formerly believed safe. Recently, outbreaks of the so-called "bird flu" have occurred, in which an avian virus is transmitted to humans through handling of birds (e.g., chickens in processing them for food) (Abbott and Pearson, 2004).

Sources of Contamination

Often contaminated water and animal feeds are the source for animals. In many instances these pathogens survive traditional preparation. For example, *E. coli* O157:H7 can persist in a rare hamburger and *Salmonella enteritidis* in an omelet or in a raw egg used for salad dressing. Bacteria can be transferred from foods intended to be properly cooked to other foods, such as when salmonella-contaminated chicken juice is on a cutting board that is then used to prepare a salad. Improper food storage can allow the growth of pathogens in food, such as *Clostridium botulinum* and *Staphalococcus aureus*.

Although commodity corn and other grain products are strictly and regularly monitored at multiple processing stages, including mills, dairy facilities, and by regulators, practically all corn or corn products contain at least tiny amounts of fungal mycotoxins. In a recent report, 363 samples of cereal-based infant food were tested, and 100 percent were found to carry various mycotoxins (Lombaert et al., 2003). Consumer illnesses, however, have not been directly attributed to these small amounts of mycotoxin exposure.

Although the potential exists for mycotoxins to reach hazardous levels, the level of monitoring makes it highly unlikely for a contamination event at hazardous levels to occur. Contaminated lots are identified and discarded, obviating the need for a recall. In a recent report for the UK Food Standards Agency, two loads of organic corn meal were prevented from being sold to consumers because of excessive levels of fumonisins, a type of mycotoxin (FSA, 2003).

On occasion, a food processing plant is a source of contamination with either a biological (e.g., *E. coli*) or nonbiological (e.g., mycotoxin) contaminant. These events may be due to the inadvertent introduction of the contaminant or to a breakdown in the usual monitoring and control systems. When recognized, such events either are corrected before consumers are exposed to a potentially hazardous food or recalled by regulatory agencies (the U.S. Department of Agriculture's Food Safety and Inspection Service and the U.S. Department of Health and Human Services, Food and Drug Administration).

Introduction of Pathogens into Food

Contaminants that are introduced early in the production process are a major problem. Introduction of contaminants can occur via contaminated animal feeds, impure water, and inadequately composted manure, or by "contamination during production and harvest, initial processing and packing, distribution, and final processing" (Tauxe, 1997). Others are introduced or enhanced in the process of food storage and preparation. Reports of incidences of bacterial food-borne illnesses in the United States between 1996 and 2001 have declined: *Yersinia* (49 percent), *Listeria* (35 percent), *Campylobacter* (27 percent), and *Salmonella* (15 percent) (Pinner et al., 2003). These declines may be due to new food safety measures that were put in place in the 1990s.

Infections with *E. coli*, however, have not shown a similar decline. The number of *E. coli* contamination events in the United States declined only between 2000 and 2001, suggesting a year-to-year variation rather than a consistent trend (Bender et al., 2004). Overall, reports of trends in meat contamination indicate that the prevalence of *E. coli* in ground beef may not have changed (FSIS, 2003).

Nutrient Deficiencies, Toxicities, and Other Nutrient Imbalances

Importantly, concerns regarding nutrient deficiencies and toxicities have been raised because of the acknowledged capability of genetic engineering to markedly change the composition of plant foods. Thus, modifications of food composition must consider the potential impact on nutrient deficiencies, toxicities, interactions, and/or other imbalances. The deletion of essential nutrients from foods or, more likely, their enhancement, has the potential of influencing the risk of nutrient deficiencies or toxicities, respectively, in the general or subsets of the population, depending on exposure patterns. In this context it should be noted that to date most nutrient toxicities are due to the addition of nutrient levels in excess of normal physiologic needs, achieved through fortification or due to the excessive consumption of nutrient supplements.

Nutrient Deficiencies and Toxicities

The concepts of nutrient deficiencies were developed several decades ago (Youmans, 1941), and have been undergoing significant change since then (Bendich, 2001). One recent conceptualization of deficient intakes is expressed by an Institute of Medicine report, that is, the "level of intake of a nutrient below which almost all healthy people can be expected, over time, to experience deficiency symptoms of a clinical, physical, or functional nature" (IOM, 1994).

This concept recognizes that single and multiple nutrient deficiencies may have multiple manifestations that are expressed at diverse levels of intake, determined by gender, age, physiological state (e.g., puberty, postmenopause, pregnancy, and lactation), genetic variability, health status, activity levels, and diet composition, and often are the result of chronically inadequate intakes rather than acute insufficiency. This conceptualization is, however, very conservative in the sense that many individuals are likely to experience signs and symptoms of deficiency before "almost" all healthy people achieve a deficiency state. Importantly, acute and chronic effects of nutrient intakes are examined in the most recent evaluation of nutrient requirement levels (IOM, 1997, 1998, 2000a, 2000b).

Although the diagnoses of specific nutrient states in individuals often are challenging, such diagnoses are relatively straightforward compared with the estimation of minimum intake levels that are required to prevent a deficiency state in an individual. Thus, the amount of a nutrient recommended to individuals to avoid deficiency is set at sufficiently high levels to minimize individual risk (to < 3 percent), that is, the Recommended Dietary Allowance, "the average daily nutrient intake level sufficient to meet the nutrient requirement of nearly all (97 to 98 percent) healthy individuals in a particular life stage and gender group" (IOM, 2001).

The concept of nutrient toxicities is relatively new. Upper tolerable levels of intakes have been set only recently by authoritative bodies (IOM, 1997, 1998, 2000a, 2000b, 2001, 2003).

Expanding Definitions of Nutrient Deficiencies and Toxicities

Other aspects of nutrient deficiencies and toxicities relevant to this discussion are the expanding definitions of nutrients and of their benefits and toxicities and the increased recognition of the roles of genetic variability in determining susceptibility to deficiency and toxicity states. The current awareness of the link between food and health reflects both relatively detailed understanding of relationships between a nutrient and a designated function or disease risk (e.g., enhanced immune function and cardiovascular disease, respectively), and less-specific associations among diet and other disease risks (e.g., cancer). These links significantly expand traditional concepts of nutrient deficiencies and toxicities (IOM, 1997, 1998, 2000a, 2000b, 2001, 2003).

Nutrient Imbalances and Interactions

Adverse health effects also may occur as a consequence of interactions among nutrients or among essential nutrients and other common food components. The underlying mechanisms are multiple. The most common are influences on uptake or excretion, changes in assimilation, and alterations in metabolism. These have downstream effects on nutrient transport and storage and on nutrient-dependent functions (IOM, 1998). Relationships between calcium and phosphorus, calcium and iron, and iron and ascorbic acid serve as examples that illustrate the complex character of problems that merit consideration in evaluations of nutrient-nutrient imbalances or interactions.

Calcium and phosphorus form complexes in chyme when the calcium to phosphate ratio falls below 0.375:1. These complexes are expected to decrease calcium bioavailability. Various clinical studies, however, have not detected decreases in calcium absorption at ratios as low as 0.08:1. Thus theory has not been supported by empirical evidence. Others point out that homeostatic compensation may account for a lack of empirical support, but that homeostatic compensation becomes progressively more difficult as calcium intakes fall below requirement levels (IOM, 1997). This is of potential concern because mean calcium intakes among vulnerable age groups in the United States are significantly below estimates of need (Alaimo et al., 1994; Johnson, 2000). Thus one must consider not only relative amounts of nutrients in assessments of interactions, but also the possible influence of the absolute intake of one or, possibly, all interacting nutrients.

To further illustrate the challenge, calcium intakes also may interfere with iron, zinc, and magnesium availability. Choosing iron for illustrative purposes, Hallberg and colleagues (1992) demonstrated a dose-response relationship between calcium intake and inhibition of iron absorption. The underlying mechanism for this interference is not clear. No one has demonstrated that iron deficiency in human populations is explained by excessive calcium intakes. On the other hand, demonstrating such relationships may be difficult among populations with low levels of iron deficiency.

The external validity of studies assessing such impacts is limited by difficulties in simultaneously controlling multiple factors with adverse or enhancing effects on iron and/or calcium chemical activity or net bioavailability. For example, the availability of non-heme iron is enhanced markedly by the presence of ascorbic acid (vitamin C). Ascorbic acid appears to enhance nonheme iron absorption linearly at ascorbic acid intakes up to 100 mg. Absorption may be improved two- to sixfold or more within this range of ascorbic acid intakes (Allen and Ahluwalia, 1997). Issues related to iron absorption become particularly relevant to populations of European extraction because of their high rates of hemosiderosis.

Nutrient interactions also may influence nutrient urinary losses. Using calcium again for illustrative purposes, the acquisition of optimal bone mass in childhood and adolescence is dependent upon several factors, such as genetic endowment, activity, and diet (Bachrach, 2001). Among the dietary factors that influence calcium excretion is sodium (Massey and Whiting, 1996). The effect of sodium on calcium retention is sufficiently large to possibly influence the acquisition of bone mass in childhood or bone loss in adulthood, especially among individuals with low levels of calcium intake. Although the renal tubular mechanism that underlies this interaction is understood incompletely, it is described sufficiently well to suggest that anticipatory reviews of nutrient physiology could arouse concerns of this nature and other analogous ones in early evaluations of new products. Thus exclusive reliance on postmarketing population studies to discover such adverse interactions or waiting until their discovery to assess biological plausibility of statistically significant relationships is not necessary.

Increased gastrointestinal losses of nutrients also occur, but mostly via interactions with food components other than nutrients. The most functionally relevant losses on a global basis relate to interactions between phytate and iron or zinc. The impact of phytates can be much broader. They also have been implicated in binding to proteins, thereby decreasing their availability as well as to calcium and starches. Binding calcium decreases its bioavailability and also may impair carbohydrate digestion since calcium ions enhance amylase activity. Phytates also bind to carbohydrates and thus may influence their bioavailability more directly (Jenkins et al., 1994). Other antinutrients also are of potential concern (see Box 5-1).

Interference with assimilation may occur because of dietary amino acid imbalances that adversely affect the biological value of a protein or of a protein

---

BOX 5-1   Potential Adverse Health Effects of Antinutrients

Antinutrients are compounds in food that inhibit the normal uptake or utilization of nutrients. In addition to those discussed in the text of this chapter, other antinutrients are of potential concern to human health. These include:

• Lectins: Lectins are compounds that typically originate in plants that can bind to epithelial surfaces and damage them through incompletely understood mechanisms. They also appear to have broader bioactivities that are as yet poorly described (Evans et al., 2002; Reynosa-Camacho et al., 2003; Santidrian et al., 2003). Nevertheless, their potential adverse effects remain an area of concern (Jenkins et al., 1994).

• Saponins: Saponins also are plant-based compounds that typically foam when added to water. They can hinder the absorption of lipid-soluble nutrients by binding bile acids and interfering with micelle formation through this and possibly other mechanisms.

• Tannins: Tannins are compounds found in nearly every plant throughout the world, and can form complexes with dietary proteins and thus impair their digestibility. Tanins also might reduce trypsin and amylase activity, thus impairing protein and carbohydrate digestion (Jenkins et al., 1994).

• Other food components with antitryptic and anti-α-amylase action, found in commonly consumed foods (e.g., cereal grains and legumes), potentially can interfere with digestion.

mixture. These are described well in the literature. These adverse effects result from inefficiencies in amino acid utilization imposed by inadequate levels of one or more indispensable amino acids. Differences in the biological values of proteins (the ratio of retained to absorbed nitrogen) are appreciated easily by contrasting the value of wheat gluten to egg protein at levels of intake that range from 0.1 g/kg body weight to 0.6 g/kg body weight in adults.

Wheat gluten's biological value ranges from 1.06 to 0.37 as intakes rise from 0.1 to 0.6 g/kg body weight and that of egg protein from 1.03 to 0.71 at intakes from 0.2 to 0.5 g/kg body weight (Inoue et al., 1974; Young et al., 1973). The ratio depends primarily on the ability of amino acid patterns of individual proteins or dietary protein mixtures to meet an organism's indispensable amino acid needs for growth and maintenance.

The potential impact of dietary amino acid balance is evident in studies of amino acid supplementation or protein mixtures. For example, the successive supplementation of wheat flour with lysine, tryptophan, methionine, threonine, isoleucine, and valine increases nitrogen retention incrementally to over three times the levels achieved without supplementation (Bressani, 1971). Studies of protein mixtures also reflect these relationships, that is, the net biological value of dietary protein depends on the proportion of protein from various sources. For example, corn and soybean have complementary amino acid patterns in the sense that although corn is relatively deficient in lysine, it supplies a relative surfeit of methionine; the opposite is true for soy.

Thus a maximum biological value is attained when corn supplies approximately 40 percent of dietary protein and soy 60 percent. The mixture's biological value falls as the proportion of either protein source falls or rises in isonitrogenous diets. Furthermore, it is possible to add a protein with an imbalanced amino acid pattern to an otherwise adequate dietary protein intake and observe adverse effects on growth rates, as some amino acids are known to cause other types of toxicities when consumed in excessive amounts, and others to do so only when their intake is excessive relative to that of a structurally similar amino acid (i.e., amino acid antagonisms with excessive intakes of leucine relative to those of isoleucine) (Harper, 1964).

Naturally Occurring Toxicants

Adverse effects can result from consuming naturally occurring toxicants in foods through several different scenarios (see Box 5-2). Some foods contain naturally occurring toxins that elicit adverse reactions only if the food is eaten in abnormal amounts. An example is the presence of cyanogenic glycosides in lima beans, cassava, and fruit pits, among other foods. Cyanide can be released from these compounds by enzymes present in the plant tissues during the storing and processing of the food or by stomach acid after the food has been ingested.

The amount of cyanide present in lima beans varies with the variety, the part

BOX 5-2   Adverse Health Effects:
Naturally Occurring Toxicants in Foods

Naturally occurring constituents of food that can cause illness (at levels that are relatively easy to reach by consumers)

- Toxicants in seafood (improper harvesting)
- Staphylococcal enterotoxins in various foods (improper storage and handling)
- Botulinal toxins in various foods (improper preservation)
- Mycotoxins in various foods (improper storage)

Unusual foods that can cause illness (at levels that are relatively easy to reach by consumers)

- Poisonous mushrooms
- Poisonous plants such as foxglove and *Senecio*
- Poisonous fish such as puffer fish

Naturally occurring constituents of food that can cause illness (with unusually high consumption)

- Cyanogenic glycosides in lima beans, cassava, and fruit pits
- Phytoestrogens in ginseng

Naturally occurring components of foods that can cause illness with usual consumption levels (only in susceptible consumers)

- Food allergens
- Lactose intolerance
- Components leading to celiac disease

Naturally occurring constituents of foods that can cause illness with usual consumption levels (only with unusual means of processing or preparation)

- Lectins in under-processed kidney beans
- Trypsin inhibitors in under-processed legumes

SOURCE: Adapted from Taylor and Hefle (2003).

of the plant, and the growing conditions. Commercial varieties of lima beans, in comparison with certain wild varieties, contain low levels of cyanogenic glycosides. However, ingesting three-fourths of a pound of lima beans may be sufficient to elicit a severe case of cyanosis (Cheeke and Shull, 1996), the result of cyanide poisoning.

Other foods contain naturally occurring toxicants that elicit adverse reactions only if the food is prepared in a manner that allows for the retention of a toxicant that is normally destroyed or discarded. For example, the lectins present in kidney beans are typically destroyed by thoroughly cooking kidney beans before eating them. Noah and colleagues (1980) reported that consumers who soaked a quantity of raw kidney beans and ate them with little or no cooking had a prompt onset of abdominal pain and bleeding. In other cases, foods may become contaminated with naturally occurring toxicants. For example, botulism and staphylococcal food poisoning are produced by bacteria, aflatoxin, and other mycotoxins by molds, and paralytic shellfish poisoning and ciguatera fish poisoning arise from aquatic algal microorganisms called dinoflagellates.

Consumer illnesses have been attributed to the very occasional presence of cucurbitacins in zucchini (Morgan and Fenwick, 1990). Cucurbitacins are thought to be formed in zucchini as a result of environmental stress, such as drought. Ingestion of these compounds may result in acute gastrointestinal illness. However, consumers often avoid eating them because they cause bitterness in the zucchini.

Opines are an example of toxicants that are generated by a bacterial pathogen (crown gall) produced in vegetables that carry the disease (discussed in Chapter 2). Opines are small carbon compounds produced by tumors that are induced by the crown gall bacteria. The opines spread throughout the plant, and therefore may be ingested when the plant is eaten, however, with unidentified effects on humans (McHughen, 2000).

Environmental or Industrial Contaminants

Toxic substances are classified in general according to their potential to cause adverse effects with acute or longer-term exposure, the organ systems affected, and types and severity of effects that they elicit. A toxin generally is defined as any endogenously produced substance that can induce a harmful response in a biologic system, causing serious injury to a specific function or organ, or producing death. The sixteenth century physician Paracelsus said that "the dose makes the poison," meaning that any substance is harmful if too much of it is ingested, and that different endpoints are associated with different dosages.

Types of Toxicity

Toxic substances often are classified according to the organ system where damage occurs, for example, to the brain and nervous system (neurotoxicity), to the liver (hepatotoxicity) and so forth. Toxic substances also may be classified by their source, effect (e.g., carcinogenic initiator or promoter), physical state (e.g., gas and liquid), chemical characteristics (e.g., proteins, heavy metals and halogenated hydrocarbons), and/or mechanism of action (e.g., cytochrome oxidase inhibition by cyanide).

Toxic effects may be local or systemic, although most often this differentiation is a matter of degree. The affected organ most likely to initiate systemic effects is the brain or, more broadly, the central nervous system. Toxins or toxicants that affect the circulatory system, blood and broader hematopoietic system, visceral organs, and the skin, in that order, also may have systemic effects, and those that affect muscle and bone generally are the least likely to have broader systemic consequences. Toxins and toxicants also may be classified by the type of damage they induce. Various examples are discussed below.

Acute Effects

Some proteins are known to be toxic (e.g., botulinum toxin, snake venoms, and plant toxins). Generally, known toxic proteins in food act via acute mechanisms at low doses (EPA, 2000; NRC, 2000). Another type of acute effect is teratogenicity. Teratogens are classified as acute toxicants because generally there is only a small window during which they can disrupt embryonic development.

Subchronic and Chronic Effects

Testing for subchronic and chronic adverse effects of specific compounds in whole foods is not a simple undertaking (see Chapters 4 and 6) and in consequence, we have very little information about the role of proteins (e.g., lectins), and other food constituents that produce such effects. Certain agents are of particular concern because of their ability to disrupt specific types of normal cellular processes and cause birth defects, mutations, and/or cancer. There is a certain amount of overlap among such agents. For example, prenatal exposure to high levels of vitamin A causes teratogenicity but, to date, has not been associated with mutagenicity.

The role of diet and cancer also remains of concern, although specific cancer-causing dietary components have not been identified conclusively to date (Doll and Peto, 1981). Regulatory oversight, through acts such as the Delaney Clause, protect consumers from agents known to be carcinogenic by not allowing them to be added to foods. Very few natural compounds in foods have been tested for their potential to produce adverse health effects (NRC, 1996), although there is some evidence for carcinogenicity of some compounds in laboratory animals (NRC, 1996).

In 1996 the National Academies suggested that these may confer risks equivalent to those associated with chemical and pesticide residues in food (NRC, 1996). For example, foods naturally contain many potential carcinogens, including hydrazines in mushrooms and caffeic acid in a range of common foods, including coffee, plums, pears, lettuce, potatoes, celery, and apples. However, there are many unanswered questions about the actual cancer risk conferred by these carcinogens and particularly how much exposure occurs in the context of their bioavailability in whole foods (as opposed to extracts that have been used for toxicity testing).

Endocrine Disruptors

Endocrine disruption is not a health endpoint, but rather a set of modes of action of chemicals involving the endocrine system. The best documented modes of action involve antiandrogen receptor activity, for example, estrogen agonist activity (e.g., genistein and other phytoestrogens), which have comparatively weaker effects than the pesticide metabolite dichlorodiphenyldichloroethylene and the thyroid antagonist, polychlorinated biphenyls. Understanding such mechanisms is important not only for improving the ability to screen and test for agents that may be harmful, but also for the development of biologically based models for dose response assessment.

Food Allergenicity

Food Allergies and Other Food Sensitivities

Foods produced through agricultural biotechnology may result in the expression of proteins new to the human diet. Some of these new proteins may induce an allergic response to sensitive members of the population. However, under typical circumstances of exposure, only a small number of the total proteins found in foods will be allergenic, or known to be associated with food sensitivities. Foods commonly found to contain allergenic proteins include peanuts, various tree nuts, dairy products, fish, shellfish, and some cereals (Metcalfe, 2003). The spectrum of food allergies and sensitivities is shown in Table 5-1.

Food Allergy

True food allergies are predominantly, though not exclusively, diseases of childhood. True food allergies are abnormal immunological responses to a particular food or food component, usually a naturally occurring protein (Bohle and Vieths, 2004). As noted in Table 5-1, allergic reactions to foods involve a variety of symptoms ranging from very mild to severe and potentially life-threatening (Bernstein et al., 2003; Sampson, 1993). True food allergies occur in an estimated 2 to 2.5 percent of Americans, or 6 to 7 million individuals (Sicherer et al., 1999; Taylor and Hefle, 2002).

TABLE 5-1   Symptoms of IgE- and Non-IgE-Mediated Food Reactions

Immune Response	Organ System	Clinical Manifestation
IgE-mediated	Skin	Urticaria/angioedema[a]
		Atopic dermatitis
	Respiratory	Rhinoconjunctivitis
		Laryngeal edema
		Asthma
	Gastrointestinal	Nausea and abdominal cramps
		Vomiting and diarrhea
		Oral allergy syndrome
		Infantile colic (rare)
		General anaphylactic shock[a]
Non-IgE-mediated	Skin	Dermatitis herpetiformis
	Contact dermatitis	
	Respiratory	Heiner's syndrome
	Gastrointestinal	Food-induced enterocolitis
		Food-induced eosinophilic proctocolitis
		Food-induced enteropathy and celiac disease
		Allergic eosinophilic gastroenteritis
		Gastroesophageal reflux
		Infantile colic (rare)

[a] Symptoms also may be provoked by the combination of ingesting specific food in conjunction with exercising but not by ingestion of the food alone or exercise alone.
SOURCE: Excerpted from Bernstein et al. (2003).

The prevalence of true food allergies is higher in children, involving an estimated 3 to 8 percent (Bock, 1987; Bock and Sampson, 2003). In one study of children, 28 percent of parents reported adverse food reactions in their children, but only 6 percent of children had food allergies (or other adverse reactions) that were documented by double-blind, placebo-controlled food challenges (Bock, 1987). As this study illustrated, adverse reactions to foods are common, but they may be over-reported by parents or over-diagnosed by physicians; not all adverse reactions to foods are allergic in nature.

Two types of immunological mechanisms are involved with true food allergies: immediate hypersensitivity reactions that are mediated by allergen-specific immunoglobulin (IgE) antibodies and delayed hypersensitivity reactions that are cell-mediated, primarily by intestinal lymphocytes and other immune cells (Taylor and Hefle, 2002). IgE-mediated food allergies elicit symptoms within a few minutes to a few hours after the offending food has been ingested. Delayed hypersensitivity reactions are associated with symptoms that occur as much as 24 to 72 hours after someone ingests the offending food.

In susceptible individuals, B cells produce allergen-specific IgE antibodies in response to the immune system's exposure to the specific allergen (Taylor and Hefle, 2002). However, IgE-mediated allergic reactions can also be provoked by exposure to allergens in pollens, mold spores, animal dander, and insect venoms. In the sensitization phase of the allergic response, the allergen-specific IgE antibodies bind to the surfaces of mast cells in various tissues and basophils in the blood.

While the sensitization phase is symptomless, subsequent exposure to the specific allergen leads to an interaction between the mast cell/basophil-bound IgE antibodies and the allergen. This interaction causes the sensitized cells to degranulate and release physiologically active mediators into the bloodstream and tissues. These mediators, including histamine, leukotrienes, and prostaglandins, are responsible for the symptoms encountered in IgE-mediated food allergies.

An IgE-mediated food allergy causes a variety of clinical manifestations (see Table 5-1), including gastrointestinal, cutaneous, and respiratory symptoms (Bock and Sampson, 2003). Oral allergy syndrome is perhaps the most mild manifestation of IgE-mediated food allergies and is associated primarily with symptoms involving the mouth and pharynx, such as oral itching, lip swelling, facial urticaria, and labial angioedema (Ortolani et al., 1988). Oral allergy syndrome is typically associated with consuming certain fresh fruits and vegetables among individuals who have been sensitized to specific environmental pollens; the implicated fruits and vegetables have allergens that cross-react with the specific pollen allergens (Ortolani et al., 1988).

The most severe manifestation of an IgE-mediated food allergy is anaphylactic shock, a rapidly developing constellation of symptoms that can be potentially fatal within minutes if not properly treated (Burks and Sampson, 1993). Food-induced systemic anaphylaxis is reportedly the leading cause of anaphylaxis admissions to emergency departments in the United States (Kemp et al., 1995; Yocum and Khan, 1994). IgE-mediated food allergies are estimated to be responsible for more than 29,000 emergency room visits and 150 to 200 deaths in the United States annually (Bock et al., 2001).

The diagnosis of food allergies can be approached in several ways. In vitro tests of food-specific IgE antibodies are available for common food allergens. A food elimination diet also can be used, in which the suspected food is eliminated from the diet for one to two weeks to test whether symptoms improve (Sampson, 1993). Open or single-blind food challenges can then be used to screen for allergic reactions to food upon reintroduction into the diet. However, the double-blind, placebo-controlled food challenge (Bock et al., 1988; Goldman et al., 1963) is necessary to confirm a food allergy when multiple food allergies are diagnosed and/or when positive responses need confirmation. Skin tests can also be used to evaluate the existence of food-specific IgE antibodies. The choice of foods used in these challenges is based on a combination of clinical history, skin tests, results of elimination diets, and clinical judgment.

Eight foods or food groups (milk, eggs, fish, shellfish, peanuts, soybeans, tree nuts [e.g., almonds, walnuts], and wheat) are responsible for more than 90 percent of all IgE-mediated food allergies on a worldwide basis (FAO, 1995). Beyond these most common allergenic foods, more than 160 other foods have been documented to cause IgE-mediated food allergies (Hefle et al., 1996). While the list of the eight most common allergenic foods or food groups is relatively consistent on a worldwide basis, other foods can be common causes of IgE-mediated food allergies in certain regions or countries as a result of cultural dietary preferences. These include buckwheat in Southeast Asia and sesame seeds in countries with principally Middle Eastern populations (Taylor et al., 2002).

Non-IgE mediated allergic reactions also encompass a variety of clinical syndromes (Table 5-1). They are expressed clinically over a period of several hours to days and are believed to have an immunologic basis. Among the most common of the non-IgE-mediated allergic reactions are various gastrointestinal syndromes occurring most commonly in early infancy and associated with milk or soybeans, common components of infant formulae (Guajardo et al., 2002; Nowak-Wegrzyn, 2003). They are believed to have an immunologic basis primarily involving either gastrointestinal eosinophils or lymphocytes (Guajardo et al., 2002; Nowak-Wegrzyn, 2003).

Food-induced enterocolitis most commonly occurs among infants allergic to cow milk or soy-based formulas. It can cause projectile vomiting and chronic diarrhea severe enough to cause dehydration (Powell, 1978). Benign eosinophilic proctocolitis also presents in the first few weeks or months of life, often in association with cow- or soy-based formula (Machida, 1994; Odze, 1995). Food protein-induced enteropathy involves protracted diarrhea, often vomiting, failure to thrive, and malabsorption of carbohydrates.

Celiac disease is an extensive inflammatory condition of the mucosa of the small intestine (Hall, 1987; Murray et al., 2003). Also known as *gluten-sensitive enteropathy* and *celiac sprue,* celiac disease is associated with sensitivity to the ingestion of the primary protein fractions of wheat, rye, barley, and related grains, the so-called gluten fraction of wheat, and related protein fractions from the other grains (Skerritt et al., 1990). The mechanism involved in celiac disease is incompletely understood, but the absorptive epithelium of the small intestine is damaged as a consequence of immune-cell-mediated inflammation, and serious nutritional deficiencies can result (Murray et al., 2003). Dermatitis herpetiformis is a related skin condition that also is associated with gluten sensitivity (Hall, 1987; Murray et al., 2003). Allergic eosinophilic gastroenteritis (Min and Metcalfe, 1991) may possibly involve food allergy. IgE may play some role in colic and gastroesophageal reflux in infants, but its role remains unclear (Kelly, 1995).

Clinical assessment of non-IgE mediated allergic reactions is similar to that for IgE-mediated reactions in terms of taking a history. However, there are no in vitro or skin tests available for the diagnosis of gastrointestinal syndromes associated with milk and soybeans in early infancy, nor is the placebo-controlled,

double-blind challenge used. The definitive diagnosis of non-IgE-me
allergy is made when objective improvements occur after the suspected offending
food is eliminated from the diet. For celiac disease, similar approaches are employed, although serological assays are frequently used (Murray et al., 2003).

Food Intolerance

In contrast to true food allergies, food intolerances involve one of several
mechanisms: anaphylactoid reactions, metabolic food disorders, or idiosyncratic
reactions (Taylor and Hefle, 2002). Anaphylactoid reactions are elicited by substances that provoke the release of mediators from mast cells and basophils without the intervention of IgE. Although this mechanism is well-described for certain adverse reactions to pharmaceuticals, evidence for the existence of
food-induced anaphylactoid reactions is largely based on individual case reports
where the mechanism is not well characterized (Taylor and Hefle, 2002).

Examples of metabolic food disorders include lactose intolerance and favism.
Favism is an intolerance to the consumption of fava beans or the inhalation of
pollen from the *Vicia faba* plant (Marquardt, 1981). Favism produces acute
hemolytic anemia in individuals who express an inherited deficiency of the enzyme erythrocyte glucose-6-phosphate dehydrogenase (G6PDH), which is critical for maintaining levels of reduced glutathione (GSH) and nicotinamide ad-
enine dinucleotide phosphate (NADPH). GSH and NADPH help protect the
erythrocyte from oxidative damage. Fava beans contain naturally occurring oxi-
dants, vicine and convicine, which are able to damage the erythrocytes of individuals with the G6PDH deficiency.

Idiosyncratic reactions refer to those adverse reactions to food experienced
by certain individuals. The mechanism underlying these responses are unknown
(Taylor and Hefle, 2002). A good example is sulfite-induced asthma (Taylor et
al., 2003). Sulfites are common food additives that are known to elicit asthmatic
reactions in sensitive individuals, particularly in situations in which exposure to
residual sulfite is comparatively high. Sulfite sensitivity reportedly affects less
than 4 percent of all asthmatics (Bush et al., 1986).

SAFETY HAZARDS IN FOOD PRODUCTS
ASSOCIATED WITH GENETIC MODIFICATION

Nature of Modification

A large number of compositional changes in foods may potentially arise from
any method of genetic modification of food. Furthermore, genetic engineering, as
previously discussed, has a higher probability of producing unanticipated changes
than some genetic modification methods, such as narrow crosses, and a lower
probability than others, such as radiation mutagenesis. Therefore, the nature of

the compositional change merits greater consideration than the method used to achieve the change, for example, the magnitude of additions or deletions of specific constituents and modifications that may result in an unintended adverse effect, such as enhanced allergenic potential. Constituents whose levels are increased could well include some of the "natural" toxins present in food, thereby enhancing the potential for adverse effects to occur with consumption of that food. Examples of deletions of specific constituents that merit consideration are those intended to enhance nutrient bioavailability by reducing barriers to absorption.

Modifications intended to enhance uptake of essential nutrients (e.g., reduction of phytic acid to improve iron or zinc bioavailability, and thus decrease the risk of iron or zinc deficiency) are particularly attractive. Paradoxically, the more effective such modifications are, the likelier are unintended effects on the bioavailability of other dietary constituents, that is, changes that increase uptake of essential trace elements also may increase the bioavailability of unwanted contaminants, such as toxic heavy metals.

Hazards that may be of concern after this type of general evaluation are toxicities, allergies, nutrient deficiencies and imbalances, risks related to nutrient displacement, and risks related to endocrine activity and diet-related chronic diseases. These categories are not exclusive. For example, although idiopathic (without known origin) reactions also are distinct possibilities, they are not discussed because, by their very nature, they are presently impossible to predict. Since many idiopathic reactions are likely genetically determined, they may be predictable in the future as genetic polymorphisms are better understood. The International Life Sciences Institute has reviewed the safety of DNA in foods (ILSI, 2002b) and has published a monograph on *Genetic Modification Technology and Food: Consumer Health and Safety* (ILSI, 2002a).

Genetic Variability

Human genetic variability likely plays an important role in adverse reactions to foods. The human genome project presents unprecedented opportunities to understand risks to diet-related disease and susceptibility to toxicities. Early haplotype (a unique combination of alleles in a specified chromosomal region) maps support the expectation that unraveling polygenic traits that likely account for a substantial portion of diet-related chronic disease risks may not be as difficult as originally projected (Gabriel et al., 2002).

The importance of genetic variability is most salient in the dominance of nutrient-related disorders for which newborns are screened routinely in much of the United States. The predominance of nutrient-related genetic screens is unlikely due to chance, and likely reflective of the predominant role that diet plays in genetic selection, a role that is understood incompletely. Nine disorders are included in current newborn screening programs; the treatment of eight of those disorders rely significantly on nutritional management: phenylketonuria, galactosemia, maple syrup urine disease, homocystinuria, biotinidase deficiency, congenital adrenal hyperplasia, cystic fibrosis, and some hemoglobinopathies (Khoury et al., 2003). The capability to screen for yet another condition of nutritional importance, celiac disease, may soon be available (Maki et al., 2003).

Two classic examples relevant to genetic variability and resulting adverse health effects related to food intake help to illustrate this source of potential concern: celiac disease and hemosiderosis. As previously noted, celiac disease is caused by gluten sensitivity. Gluten is found in wheat, barley, and rye. Hemosiderosis, a condition that results in iron overload, is due to the abnormal regulation of iron uptake. Its prevalence also has been related to the presence of pernicious anemia. Both hemosiderosis and pernicious anemia appear to be most prevalent in populations of Northern European ancestry.

It is notable that these examples were identified because of the adoption of Northern European dietary practices by other groups or became evident because of recent (in evolutionary terms) changes in European diets. Awareness of the prevalence of these conditions likely reflects the intensity with which European populations have studied themselves rather than an increased vulnerability to this type of genetic variability.

The relatively common occurrence of such genetic variants suggests that "food-relevant" genetic polymorphisms are likely to occur in other ethnic groups. This is not surprising given the role that food availability plays in defining survival and fitness. The central role of food constituents is evident in evolution and, more recently, in early studies of genetic control. What is less salient are the selective advantages associated with most traits identified to date. Furthermore, it is unlikely that such traits are limited to common dietary constituents, such as gluten, iron, and lactose.

Hereditary fructose intolerance (HFI) illustrates that such traits are not limited to historically overt conditions as those noted above. HFI demonstrates the unmasking of genetic predispositions that accompany marked changes in the food supply. As fructose has become increasingly prevalent in diets throughout the world, HFI is recognized increasingly as a disease of weaning (Cox, 2002).

Mutations of the liver enzyme fructaldolase, required for the metabolism of ingested fructose, is the cause of this condition that may result in death if unrecognized. Its phenotype is not expressed until dietary fructose levels exceed thresholds that are not well-characterized. Less salient in its acute effects but similar in its dependence on dietary challenges is the role of saturated fats in the etiology of cardiovascular disease and the marked changes in its prevalence when predisposed individuals are exposed to these common dietary fats. The recent unmasking of a genetic predisposition to type II diabetes among the Pima Indians (Kovacs et al., 2003; Lindsay et al., 2003) is a third example.

Undoubtedly, the human genome's definition will lead to the identification of other genetic variants of functional relevance to diet and it is likely that this knowledge will impact on methods to screen for potentially adverse effects and predict their functional significance. Considerations such as these can be daunting because of the likelihood of 10 to 30 million different single nucleotide polymorphisms (SNPs) in the human genome. Most anticipate that monogenic traits will be relatively easy to identify by relating specific SNPs to specific phenotypes. Of greater interest, however, are traits that are multigenic in origin.

Fortunately, recent information suggests that deciphering the basis of multigenic traits may not be as daunting as once thought. Haplotype studies in humans strongly suggest that SNPs are not distributed randomly or independently of each other, and that specific SNPs occur in defined blocks within all chromosomes. These studies also suggest that haplotype blocks vary in size, but their average size differs consistently among population groups defined on the basis of biologic, demographic, and other traits expected to influence patterns of genetic inheritance. If this is borne out by ongoing haplotype mapping efforts, assessing links between individual genotypes and diet-related diseases that are multigenic in origin appears promising (Gabriel et al., 2002).

Thus the genetic vulnerability of individuals to some compounds in foods is evident from historical and contemporary perspectives (Stover and Garza, 2002). However, the contribution that GE foods may make to this area of potential adverse health effects in unclear. Methods to predict and assess potential unintended health effects from GE foods are addressed in Chapter 6.

REFERENCES

Abbott A, Pearson H. 2004. Fear of human pandemic grows as bird flu sweeps through Asia. *Nature* 427:472–473.

Alaimo K, McDowell MA, Briefel RR, Bischof AM, Caughman CR, Loria CM, Johnson CL. 1994. Dietary intake of vitamins, minerals, and fiber of persons 2 months and over in the United States: Third National Health and Nutrition Examination Survey, Phase I, 1988–91. *Adv Data* 258:1-28.

Allen LH, Ahluwalia N. 1997. *Improving Iron Status through Diet: The Application of Knowledge Concerning Dietary Iron Bioavailability in Human Populations*. Online. The MOST Project. Available at http://www.mostproject.org/improving%20iron%20status.pdf. Accessed May 25, 2003.

Bachrach LK. 2001. Acquisition of optimal bone mass in childhood and adolescence. *Trends Endocrinol Metab* 12:22–28.

Bender JB, Smith KE, McNees AA, Rabatsky-Ehr TR, Segler SD, Hawkins MA, Spina NL, Keene WE, Kennedy MH, VanGilder TJ, Hedberg CW. 2004. Factors affecting surveillance data on *Escherichia coli* O157 infections collected from FoodNet sites, 1996–1999. *Clin Infect Dis* 38:5157–5164.

Bendich A. 2001. Beyond deficiency: New roles for vitamins and other bioactive molecules—A memorial tribute to Lawrence J. Machlin. *Nutrition* 17:787–788.

Bernstein JA, Bernstein IL, Bucchini L, Goldman LR, Hamilton RG, Lehrer S, Rubin C, Sampson HA. 2003. Clinical and laboratory investigation of allergy to genetically modified foods. *Environ Health Perspect* 111:1114–1121.

Bock SA. 1987. Prospective appraisal of complaints of adverse reactions to foods in children during the first 3 years of life. *Pediatrics* 79:682–688.

Bock SA, Sampson HA. 2003. Immediate reactions to foods in infants and children. In: Metcalfe DD, Sampson HA, Simon RA, eds. *Food Allergy: Adverse Reactions to Foods and Food Additives.* 3rd ed. Elmsford, NY: Blackwell Publishing. Pp. 121–135.

Bock SA, Sampson HA, Atkins FM, Zeiger RS, Lehrer S, Sachs M, Bush RK, Metalfe DD. 1988. Double-blind, placebo-controlled food challenge (DBPCFC) as an office procedure: A manual. *J Allergy Clin Immunol* 82:986–997.

Bock SA, Munoz-Furlong A, Sampson HA. 2001. Fatalities due to anaphylactic reactions to foods. *J Allergy Clin Immunol* 107:191–193.

Bohle B, Vieths S. 2002. Improving diagnostic tests for food allergy with recombinant allergens. *Methods* 32:292–299.

Bressani R. 1971. Amino acid supplementation of cereal grain flours tested in children, Scrimshaw NS, Altshul AM, eds. Pp. 184-2004 in *Amino Acid Fortification of Protein Foods.* Cambridge, MA: MIT Press.

Burks AW, Sampson HA. 1993. Anaphylaxis and food allergy. *Clin Rev Allergy Immunol* 17:339–360.

Bush RK, Taylor SL, Holden K, Nordlee JA, Busse WW. 1986. Prevalence of sensitivity to sulfating agents in asthmatic patients. *Am J Med* 81:816–820.

CDC (Centers for Disease Control and Prevention). 2003. Foodborne Illness. Washington, DC: Centers for Disease Control and Prevention Division of Bacterial and Mycotic Diseases. Online. Available at http://www.cdc.gov/ncidod/dbmd/diseaseinfo/foodborneinfections_g.htm. Accessed September 23, 2003.

Cheeke PR, Shull LR. 1996. Glycosides. In: Cheeke PR, Shull LR, eds. *Natural Toxicants in Feeds and Poisonous Plants.* Westport, CT: AVI Publishing. Pp. 173–234.

Cox TM. 2002. The genetic consequence of our sweet tooth. *Nat Rev Genet* 3:481–487.

Doll R, Peto R. 1981. The causes of cancer: Quantitative estimates of avoidable risks of cancer in the United States today. *J Natl Cancer Inst* 66:1191–1308.

EPA (U.S. Environmental Protection Agency). 2000. *Mammalian Toxicity Assessment Guidelines for Protein Plant Pesticides.* FIFRA Scientific Advisory Panel. Arlington, VA: EPA.

Evans RC, Fear S, Ashby D, Hackett A, Williams E, Van der Vliet M, Dunstan FD, Rhodes JM. 2002. Diet and colorectal cancer: An investigation of the lectin/galactose hypothesis. *Gastroenterology* 122:1784–1792.

FAO (Food and Agriculture Organization of the United Nations). 1995. *Report of the FAO Technical Consultation of Food Allergies.* Rome: FAO.

FSA (Food Standards Agency of the United Kingdom). 2003. Contaminated maize meal withdrawn from sale. Online. Available at http://www.food.gov.uk/news/newsarchive/maize. Accessed November 10, 2003.

FSIS (Food Safety and Inspection Service). 2003. *Microbiological Results of Raw Ground Beef Products Analyzed for* Escherichia coli *0157:H7, Calendar Year 2003.* Online. U.S. Department of Agriculture. Available at http://www.fsis.usda.gov/OPHS/ecoltest/ecpositives.htm. Accessed March 7, 2003.

Gabriel SB, Schaffner SF, Nguyen H, Moore JM, Roy J, Blumenstiel B, Higgins J, DeFelice M, Lochner A, Faggart M, Liu-Cordero SN, Rotimi C, Adeyemo A, Coooper R, Ward R, Lander ES, Daly MJ, Altshuler D. 2002. The structure of haplotype blocks in the human genome. *Science* 296:2225–2229.

Goldman AS, Anderson DW Jr, Sellers WA, Saperstein S, Kniker WT, Halpern SR. 1963. Milk allergy. I. Oral challenge with milk and isolated milk proteins in allergic children. *Pediatrics* 32:425–443.

Guajardo JR, Plotnick NM, Fende JM, Collins MH, Putnam PE, Rothenberg ME. 2002. Eosinophil-associated gastrointestinal disorders: A world-wide-web based registry. *J Pediatr* 141:576–581.

Hall RP. 1987. The pathogenesis of dermatitis herpetiformis: Recent advances. *J Am Acad Dermatol* 16:1129–1144.

Hallberg L, Rossander-Hulten L, Brune M, Gleerup A. 1992. Calcium and iron absorption: Mechanism of action and nutrition importance. *Eur J Clin Nutr* 46:317–327.

Harper AE. 1964. Amino acid toxicities and imbalances. In: Munro HN, Allison JB, eds. *Mammalian Protein Metabolism*. New York: Academic Press. Pp. 87–134.

Hefle SL, Nordlee JA, Taylor SL. 1996. Allergenic foods. *Crit Rev Food Sci Nutr* 36:S69–S89.

ILSI (International Life Sciences Institute). 2002a. *Genetic Modification Technology and Food: Consumer Health and Safety*. Washington, DC: ILSI Press. Online. Available at http://europe.ilsi.org/file/1_multipart_xF8FF_8_ILSIgmtechno.pdf. Accessed May 23, 2003.

ILSI. 2002b. *Safety Considerations of DNA in Foods*. Washington, DC: ILSI Press. Online. Available at http://europe.ilsi.org/file/RPDNAinfoods.pdf. Accessed May 23, 2003.

Inoue G, Fugita Y, Kiski K, Yamamoto S, Niiyama Y. 1974. Nutritional values of egg protein and wheat gluten in young men. *Nutr Rep Int* 10:201–211.

IOM (Institute of Medicine). 1994. *How Should the Recommended Dietary Allowances be Revised?* Washington, DC: National Academy Press.

IOM. 1997. *Dietary Reference Intakes for Calcium, Phosphorus, Magnesium, Vitamin D, and Fluoride*. Washington, DC: National Academy Press.

IOM. 1998. *Dietary Reference Intakes: A Risk Assessment Model for Establishing Upper Intake Levels for Nutrients*. Washington, DC: National Academy Press.

IOM. 2000a. *Dietary Reference Intakes for Thiamin, Riboflavin, Niacin, Vitamin B6, Folate, Vitamin B12, Pantothenic Acid, Biotin, and Choline*. Washington, DC: National Academy Press.

IOM. 2000b. Vitamin E. In: *Dietary Reference Intakes for Vitamin C, Vitamin E, Selenium, and Carotenoids*. Washington, DC: National Academy Press. Pp. 186–283.

IOM. 2001. *Dietary Reference Intakes: Applications in Dietary Assessment*. Washington, DC: National Academy Press.

IOM. 2003. *Dietary Reference Intakes: Applications in Dietary Planning*. Washington, DC: The National Academies Press.

Jenkins DJA, Wolever TMS, Jenkins AL. 1994. Diet factors affecting nutrient absorption and metabolism. In: Shils ME, Olson JA, Shike M, eds. Pp. 583-602 in *Modern Nutrition in Health and Disease*. 8th ed. Philadelphia: Lea & Febiger.

Johnson RK. 2000. Changing eating and physical activity patterns of US children. *Proc Nutr Soc* 59:295–301.

Kelly KJ, Lazenby AJ, Rowe PC, Yardley JH, Perman JA, Sampson HA. 1995. Eosinophilic esophagitis attributed to gastroesophageal reflux: Improvement with an amino acid-based formula. *Gastroenterology* 109:1503–1512.

Kemp SF, Lockey RF, Wolf BL, Lieberman P. 1995. Anaphylaxis. A review of 266 cases. *Arch Intern Med* 155:1749–1754.

Khoury MJ, McCabe LL, McCabe ER. 2003. Population screening in the age of genomic medicine. *N Engl J Med* 348:50–58.

Kovacs P, Hanson RL, Lee YH, Yang X, Kobes S, Permana PA, Bogardus C, Baier LJ. 2003. The role of insulin receptor substrate-1 gene (IRS1) in type 2 diabetes in Pima Indians. *Diabetes* 52:3005–3009.

Lindsay RS, Funahashi T, Krakoff J, Matsuzawa Y, Tanaka S, Kobes S, Bennett PH, Tataranni PA, Knowler WC, Hanson RL. 2003. Genome-wide linkage analysis of serum adiponectin in the Pima Indian population. *Diabetes* 52:2419–2425.

Lombaert GA, Pellaers P, Roscoe V, Mankotia M, Neil R, Scott PM. 2003. Mycotoxins in infant cereal foods from the Canadian retail market. *Food Addit Contam* 20:494–504.

Machida HM, Catto-Smith AG, Gall DG, Trevensen C, Scott RB. 1994. Allergic colitis in infancy: Clinical and pathologic aspects. *J Pediatr Gastroenterol Nutr* 19:22–26.

Maki M, Mustalahti K, Kokkonen J, Kulmala P, Haapalahti M, Karttunen T, Ilonen J, Laurila K, Dahlbom I, Hansson T, Höpfl P, Knip M. 2003. Prevalence of celiac disease among children in Finland. *N Engl J Med* 348:2517–2524.

Marquardt RR. 1981. Fava beans. In: Hawtin G, Webb C, eds. *Proceedings of the Fava Bean Conference*. Aleppo, Syria: Kluwer Academic Publishers/ICARDA Publications.

Massey LK, Whiting SJ. 1996. Dietary salt, urinary calcium and bone loss. *J Bone Miner Res* 11:731–736.

McHughen A. 2000. *Pandora's Picnic Basket: The Potential and Hazards of Genetically Modified Foods*. New York: Oxford University Press.

Mead PS, Slutsker L, Dietz V, McCaig LF, Bresee JS, Shapiro C, Griffin PM, Tauxe RV. 1999. Food-related illness and death in the United States. *Emerg Infect Dis* 5:607–625.

Metcalfe DD. 2003. What are the issues in addressing the allergenic potential of genetically modified foods? *Environ Health Perspect* 111:1110–1113.

Min K-U, Metcalfe DD. 1991. Eosinophilic gastroenteritis. *Immunol Allergy Clin North Am* 11:799–813.

Morgan MRA, Fenwick GR. 1990. Natural foodborne toxicants. *Lancet* 336:1492–1495.

Murray IA, Bullimore DW, Long RG. 2003. Fasting plasma nitric oxide products in celiac disease. *Eur J Gastroenterol Hepatol* 15:1091–1095.

Noah ND, Bender AE, Reaidi GB, Gilbert RJ. 1980. Food poisoning from raw red kidney beans. *Br Med J* 281:236.

Nowak-Wegrzyn A. 2003. Future approaches to food allergy. *Pediatrics* 111:1672–1680.

NRC (National Research Council). 1996. *Carcinogens and Anticarcinogens in the Human Diet: A Comparison of Naturally Occurring and Synthetic Substances*. Washington, DC: National Academy Press.

NRC. 2000. *Genetically Modified Pest-Protected Plants: Science and Regulation*. Washington, DC: National Academy Press.

Odze RD, Wershi BK, Leichtner AM, Antonioloi DA. 1995. Allergic colitis in infants. *J Pediatr* 126:163–170.

Ortolani C, Ispano M, Pastorello E, Bigi A, Ansaloni R. 1988. The oral allergy syndrome. *Ann Allergy* 61:47–52.

Pinner RW, Rebmann CA, Schuchat A, Hughes JM. 2003. Disease surveillance and the academic, clinical, and public health communities. *Emerg Infect Dis* 9:781–787.

Powell GK. 1978. Milk- and soy-induced enterocolitis of infancy: Clinical features and standardization of challenge. *J Pediatr* 93:553–560.

Reynosa-Camacho R, Gonzalez de Mejia E, Loarca-Pina G. 2003. Purification and acute toxicity of a lectin extracted from tepary bean (*Phaseolus acutifolius*). *Food Chem Toxicol* 41:21–27.

Sampson H. 1993. Adverse reactions to foods. In: Middleton E, Reed CE, Elliset EF, eds. *Allergy: Principles and Practice*. 4th ed. St. Louis: CV Mosby. Pp. 1661–1686.

Santidrian S, deMoya CC, Grant G, Gruhbeck G, Urdaneta E, Garcia M, Marzo F. 2003. Local (gut) and systemic metabolism of rats is altered by consumption of raw bean (*Phaseolus vulgaris* L var. athropurpurea). *Br J Nutr* 89:311–319.

Sicherer SH, Munoz-Furlong A, Burks AW, Sampson HA. 1999. Prevalence of peanut and tree nut allergy in the US determined by a random digit dial telephone survey. *J Allerg Clin Immunol* 103:559–562.

Skerritt JH, Devery J, Hill AS. 1990. Gluten intolerance: Chemistry, celiac-toxicity, and detection of prolamins in foods. *Cereal Foods World* 35:638–644.

Stover PJ, Garza C. 2002. Bringing individuality to public health recommendations. *J Nutr* 132(8):2476S–2480S.

Tauxe RV. 1997. Emerging foodborne diseases: An evolving public health challenge. *Emerg Infect Dis* 3:425–434.

Taylor SL, Hefle SL. 2002. Allergic reactions and food intolerances. In: Kotsonis FN, Mackey FN, eds. *Nutritional Toxicology*. 2nd ed. London: Taylor & Francis. Pp. 93–121.

Taylor SL, Hefle SL. 2003. Naturally occurring toxicants in foods. In: Cliver DO, Riemann HP, eds. *Foodborne Diseases*. 2nd ed. London: Academic Press. Pp. 193–210.

Taylor J, King RD, Altmann T, Fiehn O. 2002. Application of metabolomics to plant genotype discrimination using statistics and machine learning. *Bioinformatics* 18:S241–S248.

Taylor SL, Bush RK, Nordlee JA. 2003. Sulfites. In: Metcalfe DD, Sampson HA, Simon RA, eds. *Food Allergy: Adverse Reactions to Foods and Food Additives.* 3rd ed. Elmsford, NY: Blackwell Publishing. Pp. 324–341.

Wodicka V. 1982. *An Overview of Food Risk/Hazards.* Presented at the Food and Drug Officials Sponsored Symposium on Food Toxicology, Spring Workshop, May 23-25, New Orleans, Louisiana.

Yocum MW, Khan DA. 1994. Assessment of patients who have experienced anaphylaxis: A 3-year survey. *Mayo Clin Proc* 69:16–23.

Youmans JB. 1941. *Nutritional Deficiencies. Diagnosis and Treatment.* Philadelphia: JB Lippincott.

Young VR, Taylor YS, Rand WM, Scrimshaw NS. 1973. Protein requirements of man: Efficiency of egg protein utilization at maintenance and submaintenance levels in young men. *J Nutr* 103:1164–1174.

6

Methods for Predicting and Assessing Unintended Effects on Human Health

This chapter focuses on current and prospective approaches for predicting and assessing unintended effects on human health from genetically modified (GM) foods, including those that are genetically engineered (GE), both before and after commercialization. (For an explanation of the distinction between GM and GE foods, see Chapter 1.)

BACKGROUND

The major challenges to predicting and assessing unintended adverse consequences—such as toxicity, nutritional deficiency, and allergenicity—stem from limitations in available data as well as in current scientific knowledge. For example, information about the range of normal compositional variability, especially in plant-derived food, is very limited. This significantly constrains the ability to distinguish true compositional differences of a "new" food from the normal variation found among its antecedents.

To the extent that it cannot be determined whether the composition of a food has changed, it also cannot be predicted whether such changes have either adverse or beneficial health consequences. Even in cases where food composition changes are known, current understanding of the potential biological activity in humans for most food constituents is very limited. This becomes most evident when considering mixtures or diets consumed by human populations and then attempting to predict adverse health consequences from chronic intake of specific foods.

Thus the present state of knowledge requires relying on a range of toxicological, metabolic, and epidemiological sciences to assess the significance of un-

127

intended health effects, using both targeted and profiling approaches (see Chapter 4). Employing a combination of these approaches builds on what is known and will increase the ability to detect or even prevent unsuspected consequences.

Current approaches likely will be limited when applied to new GE foods with substantially altered composition. Consequently, a conceptual approach is presented in this chapter, grounded in the biological basis of adverse effects on human health and relying significantly on robust information regarding exposure.

Despite the power of methods suggested by this conceptual approach and their ability to identify GE foods likely to have adverse effects, it is impossible using any method to prove the lack of an unintended effect. This is particularly true given the current state of knowledge regarding the exposure patterns of U.S. populations and how single food and mixtures of food components affect health. Thus requiring proof that there is no possibility of an unintended effect is not realistic for an assessment standard.

The general conceptual approach for predicting and detecting adverse health outcomes discussed in this chapter is based on a risk assessment strategy proposed by the National Research Council (NRC, 1983) and relies on "substantial equivalence" to illustrate distinctions that may exist between foods modified by genetic engineering and those modified through traditional (non-GE) methods. This approach rests on the likelihood and functional significance of adverse outcomes of unintended or intended modifications being determined by several factors. These factors relate to the nature of the modification, such as whether it is quantitatively large or small and whether it is novel, and the characteristics of the compositional changes in question, such as dose-response outcomes and the nature and extent of likely exposures. Additionally, it considers population characteristics related to susceptibility, such as age, genetics, and nutritional status.

STAGES IN THE DEVLOPMENT OF GE FOODS

The development of a GE food involves a complex process that can be viewed as occurring in three stages: gene discovery, selection, and product advancement to commercialization. The safety of GE food should be assessed at all stages of its development (Taylor, 2001).

Starting with an initial product concept, the gene discovery stage involves screening genes from many sources and selecting those that might contribute to a marketable result. Ideally, safety assessment should begin during this early gene-selection phase by taking into account each gene's source, previous consumer exposure to the source, and whether there is a history of safe use for source material, the gene, and its specific products.

In the case of GE plants, animals, and microbes, the next stage of the developmental process is line selection. Plants, for example, progress through a variety of steps in the greenhouse and field during which the biological and agronomic equivalence of the GE crop should be compared with its traditional counterpart.

These evaluations do not specifically focus on safety assessment, but many potential products with unusual characteristics are eliminated during this stage. This elimination process enhances the likelihood that a safe product will be generated.

Finally, in the precommercialization stage for both GE plants and animals, the GE product should go through a detailed and specific safety assessment process. This process should focus on the safety of the products associated with the introduced gene and any other likely toxicological or antinutritional factors associated with the source of the novel gene and the product to which it was introduced. The safety of the GE product for both human and animal feeding purposes must be considered.

SUBSTANTIAL EQUIVALENCE AND
ITS ROLE IN SAFETY ASSESSMENT

Given the relative novelty of genetic engineering, few examples are available that involve safety assessments for GE food, especially those with substantially altered composition that are the focus of this report. The use of substantial equivalence is one approach used to illustrate distinctions that may exist between foods modified by genetic engineering compared with traditional (non-GE) methods for modifying food composition.

The concept of substantial equivalence provides a basis to plan a safety assessment designed to determine if GE foods are as safe as their traditional counterparts (FAO/WHO, 1996; IFT, 2000; OECD, 1993). It was developed in part because traditional toxicological approaches for evaluating the safety of food additives, pesticide residues, and contaminants do not work well in evaluating the safety of whole food, including GE food, because of the difficulties encountered in exaggerating the dosages of whole food in the diets of experimental animals.

The concept of substantial equivalence is frequently misinterpreted because of the mistaken perception that the determination of substantial equivalence is the end point of a safety assessment, rather than the starting point. From a safety assessment perspective, the concept of substantial equivalence merely provides a framework for focusing any safety studies on the areas of greatest potential concern. Current GE varieties of traditional crops, such as corn and soybeans, are altered very little from their traditional counterparts. Thus the safety evaluation focuses on how GE crops differ from their traditional counterparts and further assumes that the unchanged components are just as safe as the traditional counterparts (see Chapter 4). With the concept of substantial equivalence, the GE food, or food component, is compared with its traditional counterpart for such attributes as origins of genes, phenotypic characteristics, composition—including key nutrients, antinutrients, and allergens—and consumption patterns. More recently, the phrase "substantial equivalence" has evolved into "comparative safety assessment" to encompass a broader meaning that includes an analytical comparative component and a safety testing component of identified differences (Kok and Kuiper, 2003).

Three outcomes are possible from the substantial equivalence comparisons (FAO/WHO, 1996). The subsequent examples are intended to illustrate the types of adverse consequences that may occur and not to signal a "clear and inevitable danger" of food derived by deliberate genetic modifications of food by traditional or more contemporary technologies.

One possible outcome is that a GE food could be judged to be substantially equivalent to its conventional counterpart. In this case, no further safety testing would be required. However, this possibility is rather unlikely to occur in cases of GE foods that have substantially altered compositional traits compared with their conventional counterparts.

In other cases, GE foods may be judged to be substantially equivalent to their conventional counterparts except for specific differences, including the introduced traits. In this situation the safety testing likely would focus on the safety of these differences and primarily on the introduced trait or gene product. An example for this outcome would be *Bt* corn or Roundup Ready soybeans. These products are those with traits, such as enhanced nutrients or reduced toxins, that usually are expressed by single genes and that share commonality with the vast majority of currently commercialized GE plants.

Finally, the GE food could be judged not to be substantially equivalent to the conventional food or food component. Examples would include products with dramatically altered food composition, such as those aimed at improved nutritional profiles. More extensive safety assessments would be required for such products, including a more rigorous nutritional and toxicological assessment. Few products from this final category have been released into the commercial marketplace, so the nature of the safety assessment process in such cases has not yet been addressed by domestic and worldwide regulatory agencies. These safety assessments would need to be conducted in a rigorous but flexible manner, depending on the nature of the novel food product.

During the process of substantial equivalence comparisons, extensive compositional analyses are conducted on the GE crop to compare it with the conventional counterpart. The selection of an appropriate comparator or comparators is obviously a key factor in this process. Comparisons should be made to the near isogenic parental variety from which the GE food was derived, and ideally to major commercial varieties of the same food, including varieties that are important in certain parts of the world where the crop will be exported.

As noted previously, safety assessments typically focus on novel gene products and proteins, as well as any components that might be created in a GE food as a result of an enzymatic protein activity, or if the food has an effect on the metabolism of the host organism. The approaches to evaluating the safety of novel gene products are discussed later in this chapter. Limitations to assessments based on the comparisons that can be made and sampling strategies are discussed in Chapters 3 and 4.

Evaluation of Substantial Equivalence with Other Predictable Changes

Applying the concept of substantial equivalence makes it possible to focus on the intentionally introduced traits and the novel proteins produced from the inserted genes. However, other predictable differences also may be identified. The possibility of altered metabolic profiles from the introduction of novel proteins with enzymatic activity is predictable for some GE food. In the case of golden rice, enhanced levels of carotenoids are produced as a direct, intended consequence of the genetic modification (Beyer et al., 2002). While this compositional difference in golden rice is intended to be beneficial to health (Nestle, 2003), the presence of the altered levels of carotenoids must also be part of the safety assessment.

Applicability to Plant, Animal, and Microbial Organisms

The framework of substantial equivalence has also been applied to GE animals and microorganisms. Assessment of the safety of the introduced traits or novel proteins can be approached in a similar fashion, no matter what the source of the inserted gene. This approach could also be applied to the identification of compositional differences and the safety assessment of food produced by all means of genetic modification.

CURRENT SAFETY STANDARDS FOR GE FOODS

On a worldwide basis, several organizations, including the Food and Agriculture Organization of the United Nations (FAO), the World Health Organization (WHO), and the Organization for Economic Cooperation and Development, have established the background for the safety assessment of GE food (FAO/WHO, 2000; OECD, 2000). In general, these organizations have concluded that GE products are not inherently less safe than those developed by traditional breeding (IFT, 2000). Furthermore, food safety considerations are similar to those arising from the products of traditional breeding other than food additives, which are subject to different regulations and testing procedures than food products.

In the United States, the accepted standard of safety for foods produced from GE crops is the same as that for other similar food products. Under U.S. law for food additives, there must be a reasonable certainty that no harm will result from intended uses under anticipated conditions of consumption (Federal Register, 1992). There is no burden on the food manufacturer to demonstrate the safety of food products that are not food additives. However, the Food and Drug Administration (FDA) can take action against a food, including GE food, if the food presents a demonstrable safety risk.

SAFETY ASSESSMENT PRIOR TO COMMERCIALIZATION

Safety of Ingested DNA

As described in Chapter 2, genetic transfer between species has been shown to occur naturally as well as through human intervention. The deoxyribonucleic acid (DNA) present in plants, microorganisms, and animals used as food is ingested in significant quantities. Further, consumption of DNA, typically 0.1 to 1.0 g per day from food sources, is not known to be toxic (Doerfler and Schubbert, 1997). Additionally, the amount of DNA from a given GM food would likely represent less than 1/250,000 of the total amount of DNA consumed from all food sources (FAO/WHO, 2000).

However, the possibility of transferring and incorporating novel genes from GM foods into cells has been investigated in animal models, humans, and microorganisms (gut bacteria). In model experiments in which mice were orally administered high doses of bacterially derived DNA, test DNA fragments were apparently incorporated into bacterial and mouse cells (Schubbert et al., 1998). This report contrasts with others in which no transfer or only a low frequency of transfer was observed (Biosafety Clearinghouse, 2003). Furthermore, the significance of the observations of Schubbert and coworkers (1998) has been seriously questioned by others (Beever and Kemp, 2000).

As pointed out previously (WHO, 2000), the transfer of DNA from GM plants into microbial or mammalian cells, under normal circumstances of dietary exposure, would require all of the following conditions to exist:

- Relevant (potentially hazardous) genetic material in the plant DNA would have to be released from the plant cells, presumably as linear fragments.
- The released genetic material would have to survive digestion by nucleases both in the plant and in the gastrointestinal tract.
- The genetic material, once exposed to the gut, would have to compete for uptake with DNA from conventional foods.
- The recipient cells would have to be competent for transformation (uptake of the DNA), and the genetic material would have to survive enzymatic degradation by normal cellular mechanisms.
- The genetic material would have to be incorporated into the host DNA by rare enzymatic events.

The consequences of uptake of DNA by somatic mammalian cells differs from that of uptake of DNA by microorganisms, as DNA in mammalian somatic cells is not transmitted to subsequent generations, but in microbes it may be. The vast majority of known bacteria are not naturally transformable. No evidence exists for the transfer to and expression of plant or animal genes in microorganisms under natural conditions. Nielsen and coworkers (1998) did observe that plant genes could be transferred to bacteria under laboratory conditions only if homologous recombination was possible. In summary, the safety of GE products should be predicated on the characteristics of the novel protein or other product expressed by the gene, rather than on the safety of ingesting DNA or the possibility of horizontal transfer of novel genetic material to humans or gastrointestinal microorganisms.

Safety Evaluation of Marker Genes and Their Products

Products of marker genes are obvious predictable differences that should be highlighted in initial substantial equivalence comparisons. In addition to principal gene products, GE foods often contain antibiotic resistance marker genes or other marker genes that remain from the product development process. The most common antibiotic resistance marker gene expresses an enzyme called neomycin phosphotransferase II.

The safety of commonly used antibiotic resistance markers has been well-established (WHO, 1993, 2000). However, a concern exists that antibiotic resistance might be transferred from a GE plant cell to intestinal bacteria in humans. Expert groups (WHO, 1993, 2000) have concluded that there is no evidence to support that the antibiotic resistance markers currently in use pose a health risk to humans or domestic animals (WHO, 2000).

Several reasons exist for the well-established safety of antibiotic marker genes in GE foods, including:

- the lack of any evidence for the transfer of antibiotic resistance to intestinal bacteria or dietary pathogens;
- the extremely low theoretical likelihood of such transfers; and
- the use of antibiotic resistance markers for antibiotics that have limited clinical applications, such as neomycin (WHO, 2000).

While there has been an increase in the prevalence of antibiotic-resistant bacteria, it cannot be attributed to the use of antibiotic resistance markers in GE foods.

Increasingly, other methods are being employed in agricultural biotechnology that avoid the incorporation of antibiotic resistance marker genes into the commercial product. These methods include removing the antibiotic resistance marker gene after successfully transferring the desired genetic trait, or using alternative marker genes in the genetic transformation. If alternative marker genes are used, the products of these genes would also need to be evaluated for safety. Since limited experience exists with such alternative marker genes, the safety of these gene products has not yet been well established.

As previously stated, the safety of GE products should be predicated on the characteristics of the novel protein or other product expressed by the gene, rather than on the safety of ingesting DNA or the possibility of horizontal transfer of novel genetic material to humans or gastrointestinal microorganisms.

Safety Assessment of Novel Gene Products

Assessing the Potential Toxicity of GE food

Toxicological studies in animals are considered on a case-by-case basis as part of assessing the safety of GE food (Kuiper et al., 2001). The demonstration of a lack of an amino acid-sequence homology of a novel protein to known protein toxicants and rapid proteolytic degradation under simulated mammalian conditions of digestion are often deemed sufficient to presume the safety of a novel protein. However, subchronic animal toxicological studies have also been conducted on some of the novel proteins and on GE food. As noted previously, the design and interpretation of animal toxicological studies with whole food, including GE food, is challenging.

Methods exist for detecting the toxicity of chemicals in premarket evaluations. FDA has compiled a set of guidelines for toxicity testing of proposed food additives (OFAS, 2001). Similarly, the U.S. Environmental Protection Agency (EPA) has developed a number of guidelines for the toxicology assessment of pesticides, including a number relevant to the health effects of pesticides in food (OPPTS, 1996).

Present guidelines, with the exception of the oral acute toxicity test, have not been applied to the assessment of currently approved GE foods. The traditional toxicological tests may be more relevant to the next generation of GE foods that will be substantially different in composition from traditional counterparts. However, application of the tests to whole food is difficult with existing methodologies, so such testing could likely be applied only to specific unique components identified in the GE variety.

Although much testing is done in vitro, most involves feeding studies with whole animals. These studies attempt to minimize the numbers of animals used because of animal welfare concerns and because of costs. Most strategies rely on high doses of the agent under study to compensate, in part, for these limitations. Generally toxicologists first determine the maximum tolerated dose (MTD) of a substance, that is, a sufficiently high dose to cause an adverse effect, but not death. Levels close to and below the MTD are tested. This procedure is designed to maximize the tests' statistical power, but it is not without controversy since testing at the MTD induces toxic effects that can cause physiological alterations to the animals that are not relevant in the case of humans exposed at much lower levels.

In a number of rare cases, animal models have produced results that are not biologically relevant to humans at all, but generally these models have been workable. However, this approach was designed to assess conventional food additives and pesticides—not the effects of macronutrients or other food components that are difficult to isolate from whole food. Feeding food at the MTD is not feasible due to the high mass and volume of intake required, which would confound the results because of excessive caloric intake and other probable dietary imbalances.

Occasionally, subchronic toxicity tests are conducted on GE food (Kuiper et al., 2001), although such testing is not typically part of the safety assessment approach used by commercial seed companies. In these tests the GE food is fed generally to rats or mice for at least 28 days (Kuiper et al., 2001).

Feed consumption, body weight, organ weights, blood chemistries, and histopathology are among the parameters that can be assessed in such experiments. However, the complexities involved in the design of subchronic toxicity tests complicate the interpretation of the results (Kuiper et al., 2001). From a practical perspective, subchronic toxicity tests likely can only be performed with the whole GM food (or some significant component of it, such as the oil fraction) because purification of sufficient quantities of the novel protein would usually be extremely difficult.

The difficulties involved in the extraction of specific components from food for testing is illustrated by the bacterially-produced Cry9C protein in StarLink corn mentioned earlier. Moreover, traditional toxicological approaches have never been proven to have utility for testing complex mixtures, including whole food. The design of tests for food components will need to be informed by the fact that food components always occur as part of a complex mixture (see Chapter 4). The toxicity of any individual compound in food could be offset by other factors that are protective, such as those that prevent exposure by binding to dietary fiber or natural antioxidants. Likewise, the toxicity could be enhanced, for example, by facilitating absorption or by inclusion of natural substances that act via the same mechanism.

A large effort is under way to develop new microarray technologies (see Chapter 4) in order to examine patterns of DNA expression that are associated with various types of toxicity, such as immune system response, receptor biology, signal transduction, protein modification, membrane transport, growth and development, metabolism, oxidative stress, and regulation of the cell cytoskeleton (Pennie, 2002).

Assessing the Potential Acute Toxicity of Novel Proteins

Most proteins are unlikely to be acutely toxic, particularly when ingested. However, an assessment of the acute toxicity of the novel proteins introduced into GE food is one approach to preventing unintended health consequences. Nevertheless, evaluation of the acute oral toxicity of a GE food and the novel proteins it may contain should be considered. Additionally, a bioinformatics database containing the amino acid sequences of known protein toxins should be developed and maintained. This database could then be used to screen novel proteins for a sequence similar to known protein toxins. Further research will be needed to develop appropriate searching strategies to use with the database.

Currently the acute toxicities of novel proteins are evaluated as part of the overall safety assessment in certain circumstances. These experiments typically involve oral administration of high doses of the novel protein by stomach tube to either rats or mice. Because many proteins are not for the most part toxic, the evaluation of the acute toxicity of novel proteins has not been particularly revealing. Examples of this application can be found for 5-enolpyruvylshikimate-3-phosphate synthase (Harrison et al., 1996) and neomycin phosphotransferase II, a marker gene product (Fuchs et al., 1993). The assessment results from these two studies indicated that the products containing the marker genes were as safe and nutritious as their conventional counterparts.

The likelihood of the unintentional introduction of a novel, toxic protein into a GE food is extremely low, simply because proteins are rarely toxic, with a few noteworthy exceptions, such as botulinum toxins and staphylococcal entertoxins. Appropriate methods exist to assess the acute toxicity of novel proteins, and they can be implemented on a case-by-case basis as necessary.

The use of subchronic and chronic toxicity testing in animals is not currently recommended as part of the safety assessment approach. Subchronic testing may be considered in cases where the novel protein has no safe history of use (e.g., proteins with lectins that may have neurotoxic actions). The current need, however, for comparatively large quantities of material precludes the use of the purified novel protein in long-term animal toxicology studies. Thus such studies would involve the use of whole GE food, which presents challenges for experimental design and interpretation.

Assessing the Possible Allergenicity of Novel Proteins

The identification of an unanticipated allergic response to a newly introduced protein in the diet is expected to be a rare event. If such responses were not anticipated from the premarket testing phase, the identification of these rare events would depend upon medical diagnosis of the allergic response and proper attribution of this response to the GE food, as discussed in Chapter 5. However, clinical approaches to the detection of rare allergenic reactions are questionable, so the focus of current assessment approaches has been on premarket assessment.

Premarket Allergenicity Assessment

Virtually all known food allergens are proteins, so the allergenic potential of all novel proteins must be determined. In 1996 a decision-tree approach for assessing the potential allergenicity of GE food was developed that relied upon evaluating the source of the gene, the amino acid sequence homology of the newly introduced protein, the immunoreactivity of the new protein with serum immunoglobuline-E (IgE) from individuals with known allergies to the source of the transferred genetic material (specific serum screening), and the various physicochemical properties of the newly introduced protein, such as heat stability and digestive stability (Metcalfe et al., 1996).

This decision-tree approach, as modified by FAO/WHO (2000), is depicted in Figure 6-1. Additional criteria have been suggested to assess the allergenicity of GE food, including comparing the overall structural identity with known allergens, targeted serum screening, and animal models (FAO/WHO, 2001). The Codex Ad Hoc Intergovernmental Task Force on Safety Assessment of Genetically Modified Foods recommended using only information on the source of the gene, structural comparisons with known allergens (both overall structural identity of 35 percent or greater and amino acid sequence identity of eight contiguous amino acids or more), specific serum screening, and pepsin resistance because targeted serum screening and animal models have not yet been validated for use in such applications (Codex Alimentarius Commission, 2002). The Report of the Fourth Session of the Codex Ad Hoc Intergovernmental Task Force on Foods Derived from Biotechnology (FAO/WHO 2003) reviewed these assessment guidelines for inclusion in the *Draft Guideline for the Conduct of Food Safety Assessment of Foods Produced Using Recombinant-DNA Microorganisms.*

As noted, the likelihood of the unintentional introduction of an allergen into a GE food is low, but it should be evaluated in every case. Although no single test can provide complete assurance that a novel protein from a source with no history of allergenicity will not act as an allergen, the combined application of all of the approaches discussed above can provide reasonable assurance that a novel protein has a low probability of acting as one. The various tests and criteria used to evaluate the potential allergenicity of GE food have been thoroughly discussed elsewhere (FAO/WHO, 2000, 2001; Metcalfe et al., 1996; Taylor, 2002; Taylor and Hefle, 2002).

Several approaches should be considered to improve the assessment of the potential allergenicity of novel proteins. First, since structural comparison between novel proteins and known allergens are predicated on the availability of a sequence homology database of known allergens, a publicly available database, as mentioned above, should be created for use by researchers, regulators, and agricultural biotechnology companies. This database would ideally contain known allergens from all environmental sources, including food.

A scientifically rigorous approach should be used to determine which proteins should be included in this database as known allergens. Research is recommended on searching strategies to develop sound, discriminating approaches that identify potential allergens. Because pepsin resistance seems to be a characteristic of many food allergens, a need exists to standardize methods for assessing this attribute.

In those cases in which genes are obtained from known allergenic sources or the sequence comparison yields potentially significant similarities to known al-

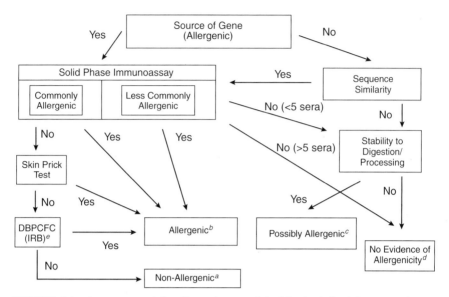

FIGURE 6-1   Assessment of the allergenic potential of foods derived from genetically modified crop plants. Adapted from a decision-tree approach developed by the International Food Biotechnology Council and Allergy and Immunology of the International Life Sciences Institute (Metcalfe et al., 1996).

[a]The combination of tests involving allergic human subjects or blood serum from such subjects would provide a high level of confidence that no major allergens were transferred. The only remaining uncertainty would be the likelihood of a minor allergen affecting a small percentage of the population allergic to the source material.

[b]Any positive results obtained in tests involving allergic human subjects or blood serum from such subjects would provide a high level of confidence that the novel protein was a potential allergen. Food containing such novel proteins would need to be labeled to protect allergic consumers.

[c]A novel protein that either has no sequence similarity to known allergens or has been derived from a less commonly allergenic source with no evidence of binding to immunoglobulin-E (IgE) from the blood serum of a few allergic individuals (less than 5), but that is stable to digestion and processing, should be considered a possible allergen. Further evaluation would be necessary to address this uncertainty. The nature of the tests would be determined on a case-by-case basis.

[d]A novel protein with no sequence similarity to known allergens and that was not stable to digestion and processing would have no evidence of allergenicity. Similarly, a novel protein expressed by a gene obtained from a less commonly allergenic source and demonstrated to have no binding with IgE from the blood serum of a small number of allergic individuals (between 5 and 14) provides no evidence of allergenicity. Stability testing may be included in these cases. However, the level of confidence based on only two decision criteria is modest. It has been suggested that other criteria should also be considered, such as the level of expression of the novel protein (FAO/WHO, 2001).

[e]DBPCFC: Double-blind Placebo-controlled Food Challenge; IRB: Institutional Review Board.

lergens, others have recommended specific serum screening as an approach to determine if the novel protein is indeed a potential allergen by virtue of its ability to bind serum IgE antibodies from humans with the specific allergy in question. However, the ability to conduct specific serum screening is limited by the lack of access to sera from individuals with well-characterized allergies to various food or environmental sources. Approaches to address this constraint should be developed. Additionally, standardized approaches for conducting specific serum screening must be developed, particularly due to the occurrence of false positive reactions in testing.

A second approach to improving methods to assess the allergenicity of novel proteins involves animal models. Animal models have been studied for IgE-mediated food allergy, but regulatory agencies have not proposed or instituted any whole animal or in vitro assays for the prediction of food allergy from novel proteins. Efforts to develop animal models for the assessment of IgE-mediated food allergenicity are under way, such as the Brown Norway rat, several mouse models, a dog model, and a swine model (Dearman et al., 2000; Ermel et al., 1997; Ito et al., 1997; Knippels and Penninks, 2003; Knippels et al., 1999; Li et al., 1999). However, none have been proven to be satisfactory (Kimber et al., 2003) due to limitations in extrapolating responses from animal models of food allergenicity to humans (Helm, 2002).

The ideal animal test model possesses several attributes: it should produce a significant amount of IgE or other Th2-specific antibody class; it should tolerate most food proteins, especially those that are known nonallergens, such as ribulose bisphosphate carboxylase; and it should develop allergen-specific antibodies on oral exposure to known food allergens.

Currently used animal models have been validated (FAO/WHO, 2001); therefore, a need exists to determine if any of these animal models can reliably discriminate between known food allergens and known nonallergens. This validation has been hindered to some extent by disagreement over which known food allergens and nonallergens should be tested. Also, sufficient quantities of the purified allergens and nonallergens have not been obtained in the quantities necessary for validation. Some debate exists about the appropriate route of exposure and whether the use of adjuvants should be permitted. More research is needed to make these determinations.

A third approach to assessing the allergenicity of novel proteins is targeted serum screening, which involves the determination of the binding of a novel protein of interest to serum IgE antibodies obtained from individuals who are allergic to materials that are broadly related to the source of the novel gene. For example, the source for a novel gene, such as a dicot plant, may not be associated with known allergies, but other dicot plants, including peanuts and various specific tree nuts, are known food allergens. The possibility exists that the novel protein may be cross-reactive with allergens from related sources.

Targeted serum screening has not been incorporated routinely into the allergenicity assessment of novel proteins from GE food. Serious concerns exist about the possibility of false positive reactions. Furthermore, if the structure of the allergens in the related species is known, the cross-reactivity should be evident from sequence homology testing. More research is needed on targeted serum screening before this approach can be recommended for routine use.

Heat processing is yet another approach that has been proposed to assess allergenicity. Empirically it has been noted that food allergens tend to survive heat processing (Metcalfe et al., 1996), so they must be comparatively heat stable. The Japanese government uses heat stability as an additional approach in allergenicity assessment (MHLW, 2000). However, this measure is not sufficient because no specific approach to heat stability assessment exists. The application of heat may alter the tertiary structure of the novel protein and thereby alter its biological functions, such as enzymatic activity, ability to kill insects (in the case of *Bt* proteins), and ability to bind IgG antibodies in animal antisera. The loss of biological functions may not completely correlate with the loss of allergenic activity, so the inability to detect a novel protein after heating does not demonstrate that the protein is no longer present in some altered form that may be allergenic.

Nutritional Evaluation of Modified Food Products

Thus far the genetic engineering of plants and animals has been aimed primarily toward enhancing agricultural productivity or improving agronomic characteristics. The development of GE food with enhanced nutritional profiles has not yet been accomplished on a commercial scale, with the exception of high-oleic acid soybeans, whose oil fraction is quite similar to olive oil (Kinney, 1996). However, several crops with enhanced nutritional characteristics are currently under development, including golden rice (Ye et al., 2000) and golden mustard (AgBiotechNet, 2000), both with enhanced levels of beta-carotene.

Such products have not been commercialized because adequate data about the safety of these products are not yet available. Certainly most of these products would not be considered substantially equivalent to their traditional counterparts. Because these types of products have not been approved for commercialization, no experience exists with respect to the adequacy of safety assessment methodology for products that are not substantially equivalent to their traditional counterparts. However, the products may be substantially equivalent to other products already in the human diet. For example, edible oils with altered fatty acid profiles (e.g., high-oleic soybeans) may be substantially equivalent to other edible oils.

Agronomic or Phenotypic Comparisons

Agronomic or phenotypic comparisons are routinely conducted as part of the line selection phase in the development of GE crops (see Chapter 3). These comparisons serve to identify varieties with altered phenotypic characteristics, and such varieties are typically abandoned. Such varieties also might have a higher likelihood of eliciting unexpected effects on human health simply because the presence of unanticipated phenotypic characteristics signifies the occurrence of unintentional compositional changes, some of which may cause adverse health effects. However, these phenotypic comparisons are rather superficial and could easily miss some varieties containing altered compositions that could adversely affect human health.

Animal Feeding Trials

In cases in which GE crops are intended to be used, in part, to feed domesticated animals, feeding trials are often conducted on cows, pigs, chickens, or sheep. The purpose of these trials is to compare the nutritional qualities of the GE crop with its conventional counterpart. Although these feeding trials are not toxicology experiments, adverse effects on the health of these animals are noted during the feeding trial. Any adverse events would indicate the possible existence of unexpected alterations in the GE crop that could adversely affect human consumers of products derived from that crop.

APPLICATION, VALIDATION, AND LIMITATIONS OF TOOLS FOR IDENTIFYING AND PREDICTING UNINTENDED EFFECTS

Hazard Identification versus Overall Risk Assessment

The safety assessment process for foods begins with hazard identification. Subsequently, other key aspects of an overall risk assessment, such as dose-response evaluation and exposure assessment are conducted, followed by risk characterization. Hazard identification in GE foods is generally based on comparisons between the GE food and its conventional counterpart to identify uniquely different components. Any potential hazards that are identified through this process that may be associated with unique components introduced into the food by genetic engineering are assessed.

When health hazards become apparent to regulatory agencies, the commercialization of GE food is likely to be stopped without any attempt to determine the overall risk using a complete risk assessment approach. As an example, the identification of a Brazil nut allergen in a genetically engineered, high-methionine soybean (see Chapter 5 for details) caused the development of this promising new crop to be abandoned (Nordlee et al., 1996). One of the challenges for the future is to incorporate the additional steps of a risk-assessment process into the safety assessment scheme; this will be particularly important in situations in which there is uncertainty regarding the hazard. A general scheme for assessing potential unintended effects is presented in Chapter 7, Figure 7-1.

Compositional Databases and Selection of Suitable Comparators

The identification of unique components in GE food is dependent upon comparisons between the composition of the GE product and a suitable comparator. Typically, these comparisons are made on the basis of proximate analysis, nutritional components, toxicants, antinutrients, and any other characterizing components. Additionally, as noted earlier, considerable focus is placed upon the unique components, usually proteins, that are produced by the inserted gene.

The selection of a suitable comparator is a pivotal and complex decision. The ideal comparator in most cases would be a near-isogenic parental variety from which the GE variety was derived. Obviously, a comparison to the near-isogenic parental variety would allow comparisons that might identify unintended changes. However, comparisons might also be needed to the most relevant comparators, that is, the commercially important varieties that are likely to be displaced in the marketplace by the GE variety. Such comparisons should be restricted to varieties that are in common use, although the identification of such varieties can be variable in different parts of the world. For crops that are likely to be exported, considerations should be made for comparisons to varieties that will be displaced in other countries, as well as varieties commonly grown in the United States.

Considerable variations can occur in the composition of a food on the basis of factors such as agronomics, environment, and natural variability. In the selection of suitable comparators, such varieties should be taken into account. Suitable comparator varieties and the GE variations might need to be grown in several different geographic areas under different environmental conditions to determine the typical ranges for key and characterizing components among these variations.

The composition of the crop can also vary among seed, leaf, stalk, and other components for each relevant variety. The compositional comparisons may be made on other portions of the plant, especially if those segments are used for animal feeds. In all such situations, suitably robust sampling strategies must be employed to obtain representative samples for analytical comparisons.

A need exists to develop compositional databases to augment identification of the range of concentrations of key characterizing components. These databases should include relevant, commercially important varieties from various parts of the world, as explained above. The databases also need to include compositional information on the edible portion of the plant and any other portions of the plant that are likely to be fed to domesticated animals. As explained in Chapter 4, such databases should include both targeted analyses and profiling analyses.

Any adverse health effect that arises from unintended compositional changes will be a consequence of the inherent toxicity of the component in question and the level of dietary exposure to that particular component. Therefore, dietary exposure should also be considered as a part of the compositional comparison. Careful attention should be paid to potential exposure levels for high-level consumers. Cultural anthropological considerations and their effect on dietary exposure should be evaluated. For example, a high proportion of the diet in some Mexican populations is corn.

Compositional comparisons should also consider the possible effects food processing may have on any unique components that might be identified in comparisons of the edible portions of the unprocessed crop. In some cases, the unintended unique components could be removed by simple processing steps, such as soaking or peeling. However, in other cases, chemical alterations would be anticipated in identified unique components as a consequence of processing operations. Again, cultural issues should be considered. Processing corn into masa or polenta would be more important in some geographic areas than in others. Processing soybeans into tofu, miso, or soy sauce would be more important in some Asian countries than in other parts of the world.

Power of Agronomic Comparisons

In the case of GE plants, the line selection stage leads to the elimination of the majority of the candidate varieties. In laboratory, greenhouse, and field trials prior to commercialization, various agronomic traits are evaluated, beginning with a rather large number of transformants. Traits such as plant height, leaf orientation, leaf color, early plant vigor, root strength, and yield may be considered depending upon the crop type. Varieties with unusual agronomic features are discarded even though these agronomic traits are not specifically linked to the safety assessment protocol. Such agronomic evaluations are important and useful, but not entirely sufficient for the identification of unintended changes.

Limitations to Toxicological Evaluation in Animal Models

Conventional toxicological tests in animals are of limited value in assessing whole food, including GE food (FAO/WHO, 2000). The amounts of food administered to animals are limited by the effects of satiety and the possibility of nutritional balance (Kuiper et al., 2001). For example, in a study of tomatoes that were genetically engineered to contain a *Bt* endotoxin, the level of lyophilized tomato powder was limited to 10 percent in the diet of rats because of the comparatively high potassium content of tomatoes (40-60 g/kg) and the possibility that higher levels of incorporation would have lead to potassium-induced renal toxicity (Noteborn and Kuiper, 1994).

In such situations it may be impossible to incorporate the GE food into the diet at levels that would be used in conventional safety assessments of individual food ingredients, that is, levels that are 100-fold or higher than those expected in human diets. Furthermore, other naturally occurring food components, such as antinutrients, could be present at much higher levels than the substances that result from the genetic engineering process, so observed abnormalities in the animals could be attributable to other factors. In cases in which the GE food is intentionally altered in its nutritional characteristics, those attributes could profoundly affect the results of animal toxicity testing.

Animal toxicology testing may be useful in some situations. The highest test dosages should be the maximum amount that can be included in a balanced animal diet, while the lowest dosage should be comparable with the expected amount in the human diet (Kuiper et al., 2001). WHO (2000) recommended that a subchronic, 90-day study in rodents with whole food should be sufficient to demonstrate the safety of long-term consumption. Longer-term studies could be considered if the results of the 90-day study indicated progressive adverse effects, such as proliferative changes in tissues (WHO, 2000). Obviously any animal toxicology studies should include carefully constructed control groups to minimize the misinterpretation of potentially confounding effects.

While the application of animal toxicology tests is of debatable value, in the case of GE food in which the novel protein is expressed at a low level, is susceptible to pepsin hydrolysis, and is not similar to known protein toxins, such testing will assume greater importance in situations in which the GE food is not substantially equivalent to its traditional counterpart and is significantly altered in its composition. In such cases, appropriate animal toxicology testing with whole food should be considered on a case-by-case basis in parallel with toxicological evaluation of individual food constituents (including novel proteins), in vitro experiments with animal and human tissues and organs, and possibly human clinical studies (Kuiper et al., 2001). Using various biomarkers to improve the sensitivity of subchronic and chronic animal toxicological tests has also been advocated (Diplock et al., 1999; Schilter et al., 1996). However, such approaches require further assessment and validation of their predictive capability.

Need for Validated Methods and Standardized Approach

The methods used in safety assessments of GE food ideally will be standardized and well-validated. Some of the current methods used for specific macronutrients (amino acids, carbohydrates, fatty acids) and micronutrients (vitamins and minerals) are reasonably well-validated and standardized. This is the case for many of the analytical procedures currently used to determine the composition of the GE food compared with their traditional counterparts (e.g., proximate analysis, nutritional profile, toxicant and antinutrient levels). As new analytical methods, such as proteomics and metabolomics, are developed (see Chapter 4), validating and standardizing them will be critical if they are to be implemented in screening procedures.

Need for Flexibility

While a core of data developed by well-validated and standardized approaches should be expected as part of the safety assessment of all GE food (including proximate analysis and nutritional profiling), many of the other comparisons between GE food and its traditional counterparts should be based upon knowledge of the inherent characteristics of the particular food and the source material providing the desirable gene. For example, characterizing antinutrients and toxicants, such as the glycoalkaloid levels in potatoes, should be compared (see also Chapter 5).

Thresholds and Adventitious Presence

The central axiom of toxicology is that the dose makes the poison. Even with respect to unintended effects on human health, the dose of the substance in question would be a major determinant of the likelihood of an adverse effect. The overall aim of the safety assessment of GE food appropriately focuses on any unique components that are intentionally produced in the GE food and do not exist in the comparator variety. The dose of any such unique component would directly influence the likelihood that this component might be associated with an adverse health effect. For metabolic components of GE food, the "threshold of regulation" concept developed for unintentional food additives should aid in the identification of those components that should be evaluated for safety (Rulis, 1992).

The threshold doses for allergic sensitization are quite low and not particularly well-defined. The possibility exists that a novel protein contained in GE food could either be or become an allergen. However, levels likely exist below which novel proteins would be unable to elicit allergic sensitization in susceptible individuals. Thus as the concentration of a potential allergen in a food decreases (dilution effect), the probability for an adverse health effect decreases as well.

EVALUATION OF POSSIBLE UNINTENDED CONSEQUENCES OF INSERTED GENES

Likelihood of Unintended Effects

All plant breeding procedures, including conventional breeding, can produce unintended effects. For example, GE soybeans altered to produce enhanced levels of the amino acid lysine (Falco et al., 1995; Hitz et al., 2002) showed an unexpected decrease in oil content, and golden rice designed to express increased levels of beta-carotene showed an unexpected increase in naturally occurring plant pigments called xanthophylls (FAO/WHO, 2000). In another example, the conventional breeding of potatoes to produce a variety with superior chipping characteristics, the Lenape variety, was developed, which had unintentionally high levels of glycoalkoloids, a class of naturally occurring toxicants typically found at low levels in commercial potato varieties (Zitnak and Johnston, 1970). This unintended effect was not discovered until after commercialization, and Lenape potatoes (see Chapter 3) were withdrawn from the market. Neither of these examples was predicted to have adverse health consequences, but they do demonstrate that unintended effects can occur.

The likelihood of unintended modifications leading to adverse outcomes is determined, to a large extent, by the method or methods used to produce intended changes. This likelihood is considered best as an incompletely understood continuum because available information does not permit the identification of clear demarcations in the likelihood of unintended modifications. Nonetheless, highly narrow crosses created by traditional breeding techniques appear to be at the low end of this putative continuum, and undirected mutagenesis by chemical radiation appears to be at the highest extreme. This probabilistic continuum is discussed in detail in Chapter 3.

Awareness of the possibility of adverse consequences will most likely result from comprehensive compositional analyses and preliminary determinations of their quantitative and functional significance (see Chapters 3 and 4). Awareness also may be raised as a consequence of more direct assessments, for example, feeding trials in humans or animals and postmarketing surveillance studies that are motivated by compositional analyses or other sources or experiences.

Importance of the Gene Discovery Stage

The initial safety assessment during the gene discovery stage is important because it highlights concerns and questions that must be effectively addressed later in the safety assessment process. Examples of health concerns that might be raised during this initial stage include the allergenicity of the source of the gene or known naturally occurring toxicants in the source of the gene. If, for example, a gene is selected from a source with a known history of allergy, such as peanuts, tree nuts, or fish, then assurance must be sought that the gene product is not the allergen from that source.

Allergenicity concerns might not be as obvious as those straightforward examples, however. As an illustrative example only, chitinase genes might be selected as a means to prevent various fungal diseases common to some crop plants. Chitinases from several plants are known allergens (Breiteneder and Ebner, 2000), so the possible cross-reactivity between the selected chitinase product and the known allergenic chitinases should be evaluated.

The *Bt* proteins used to produce various insect-resistant crops serve as a useful example of the considerations involved at the gene discovery stage. *Bt* proteins are naturally derived from *Bacillus thuringiensis*, a common soil microorganism, and more than 100 different forms of the protein are known to exist (Schnepf et al., 1998). Microbial *Bt* products have been a commercial option for insect control for several decades.

The microbial *Bt* sprays used agriculturally contain certain specific *Bt* proteins as the active insecticidal components. The particular *Bt* proteins present in these commercial insecticidal sprays have been subjected to toxicological assessment, including acute, subchronic, and chronic toxicity testing in experimental animals and oral gavage studies in humans (McClintock et al., 1995). The *Bt* proteins in these commercial products have been judged to be safe by EPA.

The various *Bt* proteins exhibit selective toxicity to specific insect targets. GE corn or maize, potatoes, and cotton have been developed that express specific *Bt* proteins within the plants. Several different *Bt* genes that express slightly different *Bt* proteins have been used or are in commercial development. The safety assessment of these *Bt* crops relies in part upon the existing history of safe use of similar products as microbial sprays. However, the various *Bt* proteins are subjected to further scrutiny, as outlined elsewhere, to supplement the existing information. In one case, a potential safety issue was identified due to further scrutiny: the comparative digestive stability of the Cry9c *Bt* protein in StarLink corn prevented its commercialization for use as a food, but not for animal feed.

In the evaluation of StarLink corn, the structure of the novel protein became a key and controversial aspect of the safety assessment. The process of extracting specific proteins from food is complex and difficult, and such extracts have not been readily available for testing. Attempts to circumvent these problems often have relied on microorganisms modified to produce the target agent. For example, in the case of the Cry9C protein, the bacterially encoded protoxin is a 129.8 kDa protein. The bacterial *cry9Ca1* gene was modified so that the inserted form in Starlink expresses a protein (Cry9c) that, among other characteristics, replaces the amino acid arginine with lysine at position 123 (APHIS, 1998). This substitution reduces the susceptibility of the active 68.7 kDa toxin to the action of trypsin. (The bacterial protoxin normally is cleaved to a 68.7 kDa active toxin whose toxicity normally is reduced by further trypsin digestion to an inactive 55 kDa fragment.) Therefore, the Cry9c protein expressed in Starlink corn is nearly identical to the bacterial toxin, but is more resistant to trypsin digestion.

For pesticide toxicity testing, EPA accepted microbially produced trypsinized Cry9c as a test substance. However, doubts have been raised about whether such equivalence should be accepted for allergenicity assessment because Cry9c may be glycosylated post-translationally in plants, but not in bacteria (Bucchini and Goldman, 2002). Allergenicity from exposure to GE foods is discussed later in this chapter.

Identification of Potential Hazards

The first step in assessing the potential of an adverse outcome is to identify suspected compositional changes and then assess their potential for adverse health effects. Adverse outcomes may be divided into two major subgroups: adverse consequences of unintended modifications that accompany presumably targeted changes, and unintended consequences of successful, highly targeted, intended modifications.

The best documented examples of an unintended modification of presumably targeted changes is the increase of psoralen in celery bred by conventional means to enhance insect resistance and breeding efforts that unexpectedly led to increased solanine levels in potatoes (Beier, 1990).

Among the best documented examples of an unintended consequence of a successful, highly targeted, intended modification achieved by transgenic techniques that resulted in a potentially harmful product is the insertion of a gene from Brazil nuts into soybeans to increase its methionine. The product included the unintended addition of a major Brazil nut allergen (Nordlee et al., 1996). This effect, while unintended, was predictable because Brazil nut allergy was well-known, and the inserted protein had to be evaluated to determine if it was a Brazil nut allergen (Nordlee et al., 1996). Further development of the product was halted once this likelihood was evident. Although field trials were conducted, there is no evidence to suggest that the Brazil nut gene crossed over into other soybean varieties.

Although dietary *trans* fatty acids were not introduced through genetic manipulation of a food, their increased consumption is among the best examples of unintended health consequences of a fully predictable compositional change in food. It not only illustrates the difficulties involved in preventing adverse health consequences of known compositional changes, but also, importantly, the ability to discover them through combined metabolic and epidemiological approaches in postmarketing phases using hypothesis-driven studies.

TOOLS FOR PREDICTING AND ASSESSING UNINTENDED EFFECTS

Genetic Analysis/Genomics

New methods of identifying proteins and genes, known as proteomics and genomics, are creating exponentially greater amounts of information about the contents of food components that have unknown relevance to human health. GE foods are typically assessed with respect to the localization and characterization of the genetic material that is inserted into the genome of the host organism. The number of copies of the inserted material is determined, as are the fidelity of the transferred DNA and the localization of those inserts with respect to other gene-coding regions and promoters. This genomic information has some value in reducing the likelihood of unanticipated effects by selecting those events that are not adjacent to or do not disrupt other genes in the genome and therefore are more likely to have unintended effects on other proteins produced by the organism. Additionally, nutrition research will be advanced by the use of nutritional genomics, which offers the potential to reduce the risk of unintended effects from exposure to certain food components and to allow for dietary planning that is focused on preventing or coping with chronic disease (Kaput and Rodriguez, 2004; Stover, 2004).

As previously discussed, many GE foods are minimally altered from their conventional counterparts. Thus novel proteins and other unique components are present at relatively low levels in these foods. Accordingly, on a dose-response basis, these novel proteins and components would not be expected to provoke unexpected adverse health effects unless they are profoundly toxic or allergenic. This situation could change considerably with the introduction of future GE products that are intentionally modified to be significantly different from their traditional counterparts. Thus greater scrutiny of such GE foods is expected during their safety assessment. However, flexibility will be required with respect to the nature of the safety assessment protocol.

Compositional Comparisons

Proximate Analysis

Proximate analysis involves determining the levels of protein, fat, carbohydrate, fiber, ash, and water in GE food. Because such determinations are relatively crude, this approach would likely identify unintended consequences only in situations in which such changes had considerable impacts on the functional and phenotypic characteristics of the food.

Nutritional Components

Nutritional analysis involves determining levels of appropriate macro- and micronutrients in the GE food. If the food is engineered specifically to enhance its nutritional characteristics, alterations in key nutrient levels would be anticipated. In other situations, changes in the nutritional composition of a GE food as compared with that of a suitable comparator would indicate the possibility of unintended effects. Such effects would not necessarily be significant in terms of human health unless the nutrient level was substantially changed in a food that served as an important dietary source of that particular nutrient. Another possibility is that genetic engineering could alter the nutrient profile or affect the bioavailability of an essential nutrient.

Endogenous Toxicants and Antinutrients

On a case-by-case basis, comparisons could also be made with respect to the levels of endogenous toxicants and antinutrients in plants, animals, and microorganisms. For example, potatoes contain naturally occurring glycoalkaloids, so glycoalkaloid levels in potatoes would typically be compared. With other foods, the identity of any such toxicants and antinutrients would be dependent upon existing knowledge. For example, soybeans contain several documented toxicants and antinutrients (e.g., phytic acid), flatulence-producing oligosaccharides (e.g., raffinose and stachyose), and trypsin inhibitors (OECD, 2001). All of these components could be assessed as part of the comparative evaluation.

Endogenous Allergens

In addition to concern about new allergens (see Chapter 5), concerns may also be expressed about endogenous allergens. Endogenous allergens are the allergenic proteins that naturally occur in specific food, for example, Ara h 1 and Ara h 2 in peanuts (Burks et al., 1991; Stanley et al., 1997) and Ber e 1 in Brazil nuts (Nordlee et al., 1996). Occasionally, safety assessment includes some consideration of the effect of the genetic engineering on the levels of endogenous allergens in the host organism. Under most circumstances, alteration in the number or levels of endogenous allergens would not be expected. In other words, both GE and conventional soybeans should be equivalently allergenic to soy-allergic consumers. If changes occurred in the levels of endogenous allergens, they would be properly characterized as unanticipated.

Glyphosate-tolerant soybeans were documented to have allergen profiles similar to those of conventional soybeans (Burks and Fuchs, 1995). The necessity of assessing GE food for altered levels of endogenous allergens is probably questionable in circumstances in which the host organism is rarely allergenic (e.g., maize) and the GE food is substantially equivalent to its conventional counterparts in all other respects. The impacts of altered levels of endogenous allergens on human health are questionable, even if they were proven to occur. For example, soybean-allergic individuals would avoid all soybeans, including GE soybeans, so that exposure to a GE soybean with higher levels of soybean allergens would have no anticipated effect on individuals already sensitized to soybeans.

An increased level of endogenous allergens might increase the likelihood of sensitization, but sensitization usually occurs after rather substantial exposure to the offending food and its allergens. Of course genetic engineering could also possibly lower the level of endogenous allergens. However, this effect would have to be quite pronounced before the GE crop could be considered hypoallergenic.

Other Characterizing Components

GE food can be compared with its conventional counterpart on the basis of any other characterizing component. For example, soybeans contain isoflavones, which may have potential health benefits, including preventing cardiovascular diseases, osteoporosis-related hip fractures, and some cancers; treating diabetes; and possibly relieving menopausal symptoms (Anderson et al., 1999; Goldwyn et al., 2000; Vedavanam et al., 1999). Thus a comparison of isoflavone levels between GE and conventional soybeans could be made. Obviously, information must exist on the range of typical levels for such characterizing components in varieties of the particular crop grown under a range of agronomic conditions before such comparisons can be meaningful.

Characteristics of Compositional Changes with Adverse Effects

Several general characteristics of food constituents are of particular relevance to early phases of identifying potential hazards. From a broad mammalian physiological perspective, the chemical structures of food constituents that are newly introduced or whose levels are altered may provide clues regarding their potential biological role in developmental and subsequent life stages through possible structure-activity relationships.

In general, the length of bioactivity of a compound and its potential for adverse effects is influenced by the stage of development during which it is first expressed (e.g., early embryonic compared with late fetal) and the number of functions the compound fulfills. In addition, if the concentration range that separates expression of a physiological role from levels that result in toxic outcomes is narrow, the compound has a higher possibility of causing adverse outcomes that may be functionally significant (Anderson et al., 2000; Vesselinovitch et al., 1979). The same is true regarding a compound's bioavailability. If a compound has high levels of gastrointestinal absorption and distribution in multiple metabolic pools and is efficient in its bioactive transformation and not efficient in its detoxification or excretion, there is a greater possibility that it will cause adverse consequences.

Other relevant factors in an agent's early evaluation are its novelty (i.e., lack of historical experience with its consumption) and allergenic potential. Obviously, potential problems are more predictable for food constituents with a long history of consumption.

Nature of Modification

The nature of compositional changes also merits consideration, for example, the magnitude of additions or deletions of specific constituents and modifications that may result in enhanced allergenic potential. It is also important to acknowledge that the most serious challenges of anticipating unintended human health consequences will be presented by components for which there is little documented knowledge. Preceding chapters have described the challenges presented by the limited information we have regarding the range of normal variation of most of the thousands of known plant constituents and their functional roles, if any, in consumers. The major exceptions are essential or nondispensable nutrients, as protocols for assessing known agents are relatively well developed (OECD, 1993, 2000; OFAS, 2001).

NEED FOR CLINICAL AND EPIDEMIOLOGICAL STUDIES

Although not a focus of this report, postmarketing studies of GE foods with substantially altered composition are also of interest because such studies often inform the selection and design of scientific approaches for assessing potential impacts, such as potential toxicity, nutritional aspects, and allergenicity, of new food prior to commercialization. Thus postmarketing studies are often vital components of an essential feedback loop that informs evaluations of food in various stages of commercialization. Similarly, timely recognition of the future potential utility of epidemiological studies can guide the premarketing development of systems to facilitate postcommercialization tracking of food or components of interest.

Clinical and Epidemiological Studies

Clinical and epidemiological studies are essential for anticipating and detecting adverse effects, identifying health outcomes, and assessing exposures. Because epidemiological approaches provide an important array of tools for anticipating and detecting adverse outcomes, there are several issues involved with interpretation of such studies that must be considered. These issues include the degree of specificity and precision for measures of exposure and outcome, study design, statistical power, potential for and control of confounding factors, analysis of effect modification, and measures of association.

Careful delineation of each of these issues is the hallmark of quality epidemiological investigations. The more tightly the exposure measurement can be defined, the stronger the interpretation of any study. Where inferences warrant interpretation beyond assessments of association, reference must be made to established criteria for causality. When toxicological studies suggest a hypothetical adverse health consequence in relation to a new GE food, postmarket epidemiological studies can be targeted to particular health consequences and, if a suspected adverse outcome is documented, aid in preventing recurrences of similar unintended effects.

Where no such suggestions arise from toxicology or other types of evaluation, routine monitoring and surveillance of the most sensitive indicators of infant health, cancer risk, cardiovascular disease risk, and other outcomes have been very valuable in detecting unanticipated problems.

Metabolic Studies

The relationship between epidemiological, toxicological, and metabolic studies can be illustrated by what happened in the usage trends and investigation of the health effects of *trans* fatty acids. Although *trans* fatty acids in food were not introduced by genetically modifying food, their introduction in food is a particularly instructive example of an unintended adverse health consequence of an intended compositional change originally designed to benefit the population at large.

It was well-established during the late 1970s that serum cholesterol levels were positively associated with coronary heart disease (CHD), suggesting that dietary factors might be responsible. This led to an increase in production of nonanimal sources of fat, namely vegetable oils, and many of these were partly hydrogenated to convert liquid oil to margarine and shortening. Concerns related to the potential for adverse consequences of partially hydrogenated oils led to studies more than three decades ago that examined the effects of partially hydrogenated fats in the diet. These studies found either modest elevations or no effects on serum cholesterol (Anderson et al., 1961; Katan, 2000; Vergriese, 1972). Metabolic studies by Katan and others in the late 1980s and 1990s showed that *trans* fatty acids had much more significant effects on overall lipoprotein patterns than were evident from changes in serum cholesterol alone, for example, increases in plasma concentrations of low-density lipoproteins and decreases in high-density lipoproteins (Aro et al., 1997; Hu and Willett, 2002; Judd et al., 1998; Mensink and Katan, 1990).

Epidemiological studies using data from several countries suggested dietary *trans* fatty acids were associated with population rates of CHD death (Kromhout et al., 1995), and in several cohort studies (Ascherio et al., 1996; Oomen et al., 2001; Pietinen et al., 1997), higher intakes of *trans* fat were associated with increased risk of CHD (Hu and Willett, 2002). The two types of evidence combined provided strong support that *trans* fatty acid intake is causally related to the risk of CHD (Willett and Ascherio, 1994).

SAFETY ASSESSMENT AFTER COMMERCIALIZATION

Postmarketing surveillance is another approach to identify unanticipated adverse health consequences from the introduction of GE food. However, postmarketing surveillance has not been used to evaluate any of the GE crops that are currently on the market, and several challenges exist to its use. First, using postmarketing surveillance presumes that the GE food will be identifiable in the marketplace, making it possible to identify consumers with exposure to that product and whose health status can then be monitored. With commodity crops such as soybeans and corn, the intermingling of GE and traditional varieties occurs on a wide scale due to shared harvesting, transportation, and storage equipment and facilities.

Consumers are often exposed to ingredients derived from GE crops, such as corn syrup or soybean oil, rather than the whole food, and some future GE food will be modified with the intent of improving the nutritional composition of the food. The incorporation of such food into the human diet presents many challenges for postmarket assessment of unintended adverse health effects. Postmarket surveillance holds considerably more promise in monitoring potential effects of

GE foods that are not substantially equivalent to their conventional counterparts and that contain significantly altered nutritional and compositional profiles.

Assessing Nutrient Profiles of Individuals

Assessing the proportion of a population at risk for a nutrient deficiency and determining the level of intake necessary to avoid deficiency in a specified proportion of a population requires that the average nutrient requirement and intake distribution be known. Methods for assessing population risk and planning for intakes of groups have been reviewed recently (IOM, 2001, 2003).

A model for establishing upper intake levels for nutrients has been developed to minimize the risk of nutrient toxicities (IOM, 1998). This model defines the Tolerable Upper Intake Level, that is, the highest level of daily nutrient intake that is likely to pose no risk of adverse health effects to almost all individuals in the general population. These levels generally are based on total nutrient intakes, regardless of source (e.g., food and nutrient supplements), and are not intended as recommended levels of intake. Intakes above the Recommended Dietary Allowance and below the Tolerable Upper Intake Level likely entail no added benefit or risk to most healthy individuals.

This model takes as its definition of an adverse health effect any "significant alteration in the structure or function of the human organism or any impairment of a physiologically important function" (Klassen et al., 1986; WHO, 1996). The model is based on two previous reports (NRC, 1983, 1994). It accommodates unique attributes of nutrients, that is, their beneficial role at lower levels, sources of variability in sensitivity, issues of bioavailability, nutrient-nutrient interactions, and other relevant factors.

An assessment of the possible toxicity of specific nutrients is also a consideration in predicting unintentional health effects of GE food. The major difference between nutrients and other potential toxins or toxicants is the certainty of nutritional benefit of intakes at and below requirement levels. Thus their elimination, unlike environmental toxicants, is never an option. Margins of safety between beneficial levels and levels that carry toxicity risks vary greatly among nutrients. Generally, those intervals are narrowest for trace elements considered essential dietary components and for inorganic and organic nutrients whose homeostatic controls to prevent excess rely significantly on storage (e.g., iron), rather than excretion (e.g., riboflavin).

Dose-Response Assessments

Predicting the functional significance of unintended modifications of food composition is likely to present special challenges because of the limited information available regarding dose-response relationships for most food constituents, even those of known functional relevance. The information base is even less satisfactory for other less well-studied food constituents. As is generally the case, dose-responses relationships depend on many factors. Among these are the characteristics of the compounds of interest, the age and physiologic state of the host, and the vehicle used to introduce the agent.

Less understood is the role of human genetic variability in determining dose-response relationships for individuals and defined population groups. This is likely to become increasingly prominent in assessments of nutrient deficiencies, toxicities of nutrients and other food constituents, and other types of adverse outcomes.

Exposure Assessment

Epidemiological studies are characterized by investigating the associations between exposures and health outcomes in defined population groups. These may include physical measures, such as height, weight, skinfold thickness, and blood pressure, and biochemical measures, such as serum lipids and serum vitamin levels. Some of these can be considered as surrogate or intermediate outcomes. Other endpoints may reflect specific conditions or diseases, such as cancer and cardiovascular disease.

Whether the test of an association between an exposure and a surrogate or health outcome will reliably indicate that there is an association between the exposure and, for example, cancer, is important to know. The conditions necessary for this inference to be valid include that the surrogate endpoint is associated with the outcome (e.g., cancer or heart disease), that the surrogate or health outcome is associated with the exposure, and that the surrogate endpoint mediates the association between the exposure and health outcome (Schatzkin and Gail, 2002).

Epidemiological studies based on surrogate or intermediate outcomes should take advantage of causal pathway diagrams and can usefully pave the way for later definitive studies (Schatzkin and Gail, 2002). Exposure assessment is key, however, to understanding putative relationships between targeted outcomes and the agent of interest.

Observational studies are often the only option in evaluating such relationships. To be useful, these must be designed as carefully as possible to mimic the gold standard for inferring causality from associations, namely the randomized controlled trial. Nonetheless, if the exposure is ascertainable or the dose can be quantified, a range of other epidemiological study designs is possible, such as ecological, feeding, occupational observational, and population- and cohort-based observational studies.

Study designs that use exposure and outcome information at the individual level have quite powerful inferential capacity. If exposures are not identifiable because GE food is mixed with conventional food to such an extent that distinctions are completely blurred, it may be that exposure can only be estimated by region and time (see Box 6-1). These are labeled group or ecological-level stud-

---

**BOX 6-1   Mixing Seed Makes Tracing Exposure to
GE Food Difficult**

Epidemiological studies of populations that consume GE food offer the potential to provide valuable information about unintended health effects, but how are such studies influenced when a given population has been exposed to mixtures containing conventionally bred, as well as GE, food? In the Rio Grande do Sul province of Brazil, it was estimated that up to 60 percent of the 2002 soybean crop might have been derived from genetically engineered seeds (nicknamed "Maradona beans") transported illegally from Argentina (Copple, 2002). In that same year, the Brazilian government reported that roughly 15 to 30 percent of the nation's soybean crop was genetically engineered (Sissell, 2002). The implications here are apparent: as crops such as soybean routinely consist of conventionally bred organisms mixed with genetically engineered organisms, it will become increasingly difficult to determine consumption patterns that illustrate who is consuming what and in what quantity. Equally problematic is the international dimension of tracing exposures to food such as that derived from Brazilian soybeans. As the producer of 25 percent of the world's soybean crop, exported seed that has been mixed with genetically engineered seed complicates efforts to trace supranational exposures to GE food. Although the Brazilian government recently decided to suspend its ban on GE crops indefinitely (Smith, 2003), the mixing of seed remains a challenge for tracing exposures to these foods.

---

ies, and are generally used for hypothesis generation. If associations are found, more targeted studies of a toxicological or epidemiological nature should be performed.

Methods for characterizing actual exposures to GE food as part of dietary intake will depend on whether or not individual exposure can be identified and estimated. Surrogate exposure assessment should also be considered. Exposure likely will need to be characterized beyond "ever/never" to recent dose or cumulative exposure. Assessments may be made in population studies at either the individual level or the group or geographical level.

Individual Level: Traditional Methods of Dietary Intake Assessment

The three forms of dietary intake assessment most commonly used in epidemiological studies are 24-hour recalls, food records, and food-frequency questionnaires (Willett, 1998). These assessment tools ascertain intakes from different food items and convert information to intake using appropriate nutrient databases. They can be used to assess intake of whole food also. Each method has attendant advantages and disadvantages that have been extensively documented (Bingham, 1991, 2002; Kipnis et al., 2002; Thompson and Byers, 1994; Willett, 1998). The large measurement error associated with these methods continues to plague epidemiological studies of food and food components.

The accuracy and usefulness of dietary intake assessment tools, however, ultimately depends on the accuracy of their supporting food composition databases (LSRO, 1995). Changes in breeding practices and improvements in biotechnology have occurred since these databases were established, resulting in genetic changes in plants and animals—highlighting the need for an ongoing process for updating nutrient composition databases to ensure accurate data on foods consumed.

Two large federal surveys have been conducted to determine dietary behaviors of U.S. consumers: the Continuing Survey of Food Intake by Individuals (CSFII) and the National Health and Nutrition Examination Survey (NHANES). The CSFII was conducted from 1994 to 1996 and in 1998. It provided information on two-day food and nutrient intakes by 20,607 individuals of all ages, drawn from a nationally representative sample of the U.S. population. This survey allowed certain inferences to be made regarding differential consumptions of various foods by children versus adults, by region of the country, and for major racial and ethnic groups.

NHANES is conducted to assess the health and nutritional status of the U.S. population and includes dietary information useful to exposure assessment. NHANES and CSFII were combined into a single survey in 1998 and became a continuous, annual survey program in 1999. A national sample of 6,000 people of all ages is taken each year. Dietary intake data vary across the different versions of NHANES. Often, assessing commodities intake requires using recipes to "translate" intake data for foods identified by survey participants (e.g., muffins) to agricultural commodities (e.g., wheat, soybeans, blueberries, and eggs).

A more targeted approach using repeat cross-sectional surveys, plus nested cohort studies, is an alternative strategy that can be designed specifically for a postmarketing surveillance study (Kristal et al., 1998). An assessment plan for surveillance of exposures and health effects would include population-based, cross-sectional samples conducted at regular intervals, with baseline assessments ideally conducted prior to marketing. Based on previous studies, the highest response rates to cross-sectional surveys are expected from approaches that rely on random digit dialing and make short time demands on respondents (no more than 10-15 minutes). This strategy is limited to households with a telephone, typically more than 95 percent of those in the United States.

Alternative population-based strategies include direct mail, limited to people on mailing lists that may be purchased (e.g., those based on registered voters or licensed drivers), and residence-based, multistage sampling. Within the group of respondents to the baseline cross-sectional survey, volunteers can be sought for more detailed dietary assessment. Within these volunteers, a small cohort would be selected for repeat assessments at the interval assessment points. The detailed dietary assessments could include multiple 24-hour recalls and serum or buccal cell nutrient assessment.

Marketing Data

It is common practice in grocery stores to have member discounts and frequent shopper discount cards. The data collected from these cards provide useful information to both processors and retailers who can tailor production or stock store shelves to meet specific local demands or to conduct targeted marketing. With consumer consent, these data have the potential for inclusion in epidemiological studies as a measure of exposure to specific food types. Investigators could approach grocery chains in several geographical areas to obtain consent and request that certain frequent shoppers be identified from their database. Simple consent forms would then be available at the checkout stand and a random sample of frequent shoppers could be invited to participate in surveys. Researchers also could contact participants to obtain additional health and other information. The sampling frame could be limited to consenting frequent shoppers in a way that allows linkage to individual-level exposure assessments and appropriate follow-up.

Group or Geographical Studies

In the absence of identifiable individual level exposure data, aggregate population analyses known as ecological studies can be used with appropriate caution as to their interpretation. This approach may be the only option if GE products are not identifiable by chemical, biochemical, biological, or genetic analyses. In this case, the approach would rely on the ability and willingness of marketing companies to identify geographical centers where their GE products were distributed widely and other centers where they were not available. Health outcome data could be collected from random samples of residents of the contrasting areas and averaged within each center, or mortality and morbidity data could be routinely collected for the regions corresponding to the contrasting centers.

An example of this potential is provided by Linola, an edible variety of linseed that was developed in Australia (Dribneki et al., 1996; Green, 1994). This crop was distributed only to Canada, the United Kingdom, and some other European countries (IENICA, 2003). Because there are multiple areas where Linola is distributed, health outcome data could be obtained as described above. Outcomes in Linola-using areas could be compared with those from non-Linola-using areas, adjusting for age distributions, other sources of oils, and other confounding variables. Differences would be tentatively attributed to Linola. It should be noted that this is a theoretical example only, and it should not be inferred that any health risks have been identified following the use of Linola.

It has been well-documented that aggregate-level information on only exposure and outcome cannot be used to rule out confounding information. Nonetheless, with the addition of some individual-level data on exposure, the inferential power of these studies can be improved (J. Wakefield, University of Washington, personal communication, 2003). The use of multiple cross-sectional surveys in this context of adding some individual-level exposure data to an ecological study would also be appropriate if it is possible to do so (Kristal et al., 1998).

Marketing data may also be gathered in a nonidentifiable way by aggregating sales information at the grocery store level. Communities served by specific grocery stores could be monitored for health effects reported at the smallest aggregate population level. Information from the Census could also be linked at this level.

The use of routinely collected federal outcome data to detect a change in frequency over time of a particular health outcome has been successful as a starting point in evaluating a potential environmental hazard on several occasions, such as the Chernobyl accident and a possible increase in trisomy 21 or neural tube defect congenital anomalies (Little, 1993). As previously stated, these designs are used most often for hypothesis generation. This approach also may be used when exposure to specific GE products is identifiable, but no information is available about specific health effects, whether intended or unintended, that may be associated with its consumption.

Linking Exposure to Outcomes

Specific Studies

If there are hypothetical health risks based on animal studies or premarket volunteer testing that are associated with GE food or food products, epidemiological studies can be designed specifically to evaluate these possible risks. The most efficient design in such instances is the population-based, case-control study. In practice this is rarely the first step in the investigation of a possible link between a new exposure and health outcomes. To illustrate its utility, it was the method used in a large study of workplace exposures and esophageal cancer (Parent et al., 2000). The study was within a defined geographical area and all historically confirmed cases of esophageal cancer in the male population aged 35 to 70 years were identified. Population controls were obtained using random digit dialing and frequency-matched to the cases based on age. A large number of potential occupational exposures were examined, and exposure to sulphuric acid appeared to be consistently related to esophageal cancer (Parent et al., 2000).

Other study methods that have been used include the repeat cross-sectional design with nested cohort that has been described previously. Early results from the study of Olestra, a fat substitute, found significant reductions in plasma concentration of vitamin E and beta-carotene (Westrate and van het Hof, 1995). Subsequently, FDA required the addition of fat-soluble vitamins to foods made with Olestra (FDA, 1996). Another study, examining a cohort of adults before and immediately following the first introduction of Olestra into the marketplace, found no evidence of decreases in serum carotenoids or fat-soluble vitamins, and there was even some suggestion of increases in vitamin K levels associated with Olestra consumption (Thornquist et al., 2000).

In some instances, a randomized controlled trial may be indicated, as was done as part of the safety evaluation of the artificial sweetener aspartame (Leon et al., 1989).The trial was of aspartame in capsule form, delivering a dose that would be equivalent to consuming 10 L of an aspartame-sweetened beverage daily. The study was placebo-controlled and lasted for 24 weeks. Vital signs, blood levels of amino acids and methanol, urinary excretion of formate, and self-reported symptoms were ascertained at intermediate points in the study and at the end. No statistically significant differences between the two groups were found (Leon et al., 1989). Another small trial among seizure-prone individuals failed to find an increase in seizures associated with aspartame compared with placebo during five days of intense monitoring (Rowan et al., 1995).

In the unlikely event that exposure to GE food is geographically limited, approaches similar to those used to study food outbreaks and other geographically concentrated potential hazards may be useful. For example, a 20-year follow-up study was conducted of mortality in an area exposed to dioxin as a result of the industrial accident in Seveso, Italy (Bertazzi et al., 2001). Cohorts of male residents based on four zones of varying contamination level were established, contrasting exposures were corroborated via lipid-adjusted plasma concentrations of dioxin, and mortality experience was compared. Although no specific cancers were found to be causally associated with dioxin exposure, an increased risk for all cancers and possibly cancers of the rectum and lung were reported (Bertazzi et al., 2001).

The challenges that such approaches may pose should not be underestimated. Prospective and retrospective, population-based, and observational studies generally document significant associations of specific dietary and other related lifestyle patterns with good or poor health outcomes (Rimm et al., 1993; Stampfer et al., 1993). Such studies generally link various dietary patterns, supplement use, and physical activity patterns to diet-related diseases, such as cardiovascular disease, diabetes, hypertension, and selected cancers, and conversely their avoidance or significant deferment. Yet attempts to achieve similar results in controlled interventions, based on single or groups of nutrients, have been largely unsuccessful.

Negative trials targeting beta-carotene to prevent lung cancer among smokers and vitamin E to prevent cardiovascular disease are two salient examples (GISSI Trial Group, 1999; Yusuf et al., 2000). For the most part, the active agent responsible for putative benefits has not been identified, thus prescribing a level of intake to avoid a functional deficiency state, or identifying a specific food constituent that should be conserved in the diet or bred at higher levels through genetic manipulations of food, has not been possible.

Trials that rely on food-based interventions have been mixed. Two relatively recent studies do not support a positive benefit of diets high in fiber content on colon cancer (Alberts et al., 2000; Schatzkin et al., 2000). Both, however, were criticized as inconclusive because of the limited duration and the doses used (Story and Savaiano, 2001). Peters and colleagues (2003) compared results from food frequency questionnaires in a case-control study and found that subjects with the highest amounts of fiber in their diets had the lowest incidences of colon adenomas compared with subjects who had the least amount of fiber in their diets. Another study (Bingham et al., 2003) used a prospective cohort to compare the dietary habits across 10 countries. Their findings indicate that people who ate the most fiber had a significantly lower incidence of colorectal cancer that those who ate the least fiber.

In contrast, the DASH (Dietary Approaches to Stop Hypertension) trials are viewed as supporting a link between diet and hypertension. Svetkey and coworkers (1999) concluded that a diet rich in fruits and vegetables and including low-fat dairy products was effective in reducing blood pressure in the general population and in those with stage I hypertension. A key observation was that neither sodium restriction nor weight loss was required to attain the putative antihypertensive benefit, components that many felt were key to efficacy. Saks and colleagues (2001) extended these findings. They concluded that sodium restriction superimposed on the DASH diet was yet more effective in reducing blood pressure, especially among African Americans, who appear to be especially vulnerable to this condition.

Thus anticipating risks due to unconventional deficiency states (e.g., increased risk of cancer) or unconventional toxicity states (e.g., hypertension) appears unlikely using current methodologies, except in the cases a few food components (e.g., sodium and saturated fats). This situation is likely to persist until more sensitive methods become available for identifying mechanistically linked biomarkers that are expressed before clinical conditions become evident, or until a better understanding is obtained of the active agents in food that increase or decrease such risks. Inadvertent changes in the content of these active agents in food, in theory, could have long-term risks or benefits. This situation is particularly challenging given historical difficulties in evaluating the health effects of such dietary components as sodium, saturated and *trans* fatty acids, and antioxidants.

Evaluating Adverse Human Health Impacts

If there are some unexplained adverse health clusters that are picked up by regional comparisons or examination of time trends, it may be advisable to conduct targeted investigations into whether changes in the food supply might be associated with the clusters. Sources of foods would be subject to scrutiny, including GE foods. As part of this process, the potential for simultaneous consideration of a large amount of microconstituents such as found in nutrient profiles or microarray studies, occurs. Indeed, the likelihood of adverse effects and the possibility for these to be functionally relevant will increase with greater net addition or deletion of targeted constituents, higher intake of modified food by vulnerable groups, and more widespread availability and accessibility of the modified food available to those groups. Nevertheless, predicting and assessing these effects, at the current state of the science, will be difficult.

As new profiling technologies (e.g., microarray studies) begin to provide much greater amounts of data presumably linking specific dietary constituents, dietary patterns, and health outcomes, new statistical methods will be needed especially to control for possibly high rates of positive or negative associations. There are approaches that are less conservative than the original Bonferroni method of correction for multiple testing of statistical significance (Efron and Tibshirani, 2002; Hochberg, 1988). One model-based method uses a regression-based Bonferroni adjustment (Thomas et al., 2001).

Another nonmodel-based approach introduces an alternative measure, denoted number of false discoveries (NFD), that is less conservative than the adapted Bonferroni adjustment. With NFD, the criteria used to accept associations as true from among a large number of tests is determined in advance to control the total number of associations that are false. The NFD can be estimated in many ways, including by permutation tests (Xu et al., 2002). Other model-based approaches are available as well, including the use of rank transformation procedures to reduce heteroscedasticity and outlier effects, the formulation of a model that approximates the relationship between microarray or nutrient profiles and experimental factors, and the use of least squares and estimating equation techniques to estimate the parameters in the model (Xu et al., 2002).

The exponential increase in available genomic data has further highlighted the need to investigate nonclassical statistical methods, such as multicomponent, model-based methods and Bayesian approaches, to evaluate them. Model-based methods impose structure on the data in order to make inferences that cannot be made without these structures. The structures arise from explicit consideration of the exposure measures and sources of variability. Model-based methods may alternatively make some estimates more precise by using information contained in the model structure. Complex models may not be able to be evaluated using classical estimation procedures, and the complexity of models under consideration can be extended using Bayesian approaches. Thus in this context, Bayesian approaches for model-based methods refers to an estimation procedure that makes complex models of health outcomes in relation to multiple exposure measures amenable to evaluation.

In summary, new statistical methods are emerging to address the complex pathways of basic physiology and disease etiology (Boguski and McIntosh, 2003).

They hold the potential for managing the increasingly large amount of information (Thomas et al., 2001) and highlighting where functional, biological, and health relevance might lie (Thomas et al., 2001; Xu et al., 2002).

Consumer's Physiological Characteristics

Nutritional needs and toxic vulnerabilities are known to vary with age, general health status, gender, and other similar characteristics, and to be influenced by physiologic states, such as pregnancy and lactation. Thus the more targeted the likely consumption of GM food, the easier it should be to assess the influence of these types of characteristics on the possibility of adverse effects and their probable functional significance.

Life stage is among the most obvious characteristics that merit consideration. The interplay between physiological characteristics and dietary behaviors is evident in children. Relative to body weight, children consume more water, food, and air than do adults. Although consequences of this relationship often depend on metabolic competencies, generally it increases vulnerability to food-borne problems. This interplay is evident among 1- to 5-year-old children who consume three to four times more food on a weight basis than do adults.

Such physiologically imposed requirements intersect with specific eating patterns in the United States such that the average 1-year-old child consumes over 20 times per body weight more apple juice and two to seven times more bananas and carrots than do adults. Marked variability in intakes of specific food is likely to increase as niche markets and specialty foods proliferate and ethnic diversity increases.

Other characteristics that merit deliberate review are the general health, economic dependencies, and prevalence of relevant genetic variants of targeted populations. The first two are of particular concern when considering populations beset by marked levels of undernutrition or a high prevalence of diseases that adversely influence organ systems responsible for physiological functions of the types discussed in preceding sections. These concerns are particularly relevant when assessments are performed in populations free of such conditions and are extrapolated to populations burdened by them.

Classic studies by Keys and colleagues (1950) document the general deterioration of multiple physiological systems during starvation. These studies were controlled carefully to elucidate the consequences of energy insufficiency. The combined ill effects of single or multiple micronutrient deficiencies superimposed on energy insufficiency are not well understood beyond the widespread recognition that functional reserves in individuals so affected are shallow and, thus, they are vulnerable to challenges that normal individuals can easily handle.

Similarly, diseases with multisystem consequences also merit concern. The most obvious contemporary example is human immunodeficiency virus. It is of interest to this discussion as an example of a devastating disease with multisystem involvement and of very uneven global distribution. Safety evaluations of products likely to be consumed by populations with high infection rates require deliberate considerations relevant to immunological, gastrointestinal, and other physiological functions.

Economic constraints that impose overwhelming dependencies on specific food staples raise concerns related to the extent and duration of exposure. The majority of the world's population depends daily on a limited number of foods. The most notable are corn, rice, wheat, and cassava. Changes in these products will impact total dietary intakes to a level proportionate to their dietary dominance. Under conditions of dietary monotony, three factors increase the possibility of adverse consequences and the likelihood of their functional significance: intakes of the modified product are high; exposure is prolonged and persists through most, if not all, of the life course; and the proportion of those exposed is nearly universal. The latter is particularly relevant if genetic variability significantly influences adverse responses to the agent's consumption.

Measurement Error and Confounders

Considerable error may be associated with exposure assessments, resulting in exposure misclassification. Such misclassification usually results in an estimate of the association between exposure and outcome that is closer to the null than to the truth (Kelsey, 1996; also see Rothman and Greenland, 1998). In targeted studies at the individual level, the estimate can be improved with the use of a validation substudy that is internal to the main study (Spiegelman et al., 2001). In the validation substudy, a more detailed dietary intake assessment would be performed using repeat measures. For example, these might include multiple blood draws for biomarker assessments.

When the validation study is large enough, with more than 340 participants, an internal estimate of the association of exposure with the outcome can be obtained and adjusted for misclassification (Spiegelman et al., 2001). A balanced design for the validation study in which an equal number of people are sampled in the four groups defined by the imperfect exposure measure and the outcome variable has been shown to perform well (Holcroft and Spiegelman, 1999). The consequences of exposure misclassification on exposure assessment are greater for exposures that are rare than for exposures that are common.

Self-selection issues have plagued epidemiological investigations that rely on observational studies. Exposure to a factor might be higher or lower because of association with a personal characteristic that is also associated with the health outcome. Health-conscious people may be healthier than others and are likely to follow healthy behaviors. Associations between the exposure and health effect might then be due to the confounder, which may or may not be measured. This kind of pitfall is well known, and methods to minimize confounding in the study design and cautions in inferences made from observational studies are standard practice (Kelsey, 1996; Lilienfeld, 1994).

Health Monitoring

In specifically designed epidemiological studies or in surveillance studies using routinely collected data, the health outcomes studied fall into four broad classes: mortality (disease specific), incidence (disease specific), integrated measures of health (e.g., health related quality of life or quality adjusted life years), and intermediate health outcomes (e.g., serum cholesterol and other lipid markers). In specific studies, due care is taken to obtain an unbiased assessment of health outcome that is complete and such that the health effects estimated are reproducible. Systems that are in place for routine surveillance include the national health surveys discussed earlier.

A key factor in postmarket surveillance is the collection and analysis of adverse effects reports. Such reports usually come from health care providers and give an indication of possible adverse effects of an agent so that investigators can determine if there is a pattern of such effects that may be causal. This was done with birth defect diagnoses and exposure to clarithromycin in which pharmacy and hospital claims were monitored (Drinkard et al., 2000); with medical reports associated with aspartame use (Butchko et al., 1994; Tollefson and Barnard, 1992); and with adverse events monitoring with respect to Olestra (Slough et al., 2001).

Birth defect monitoring systems are maintained by most states; the completeness of the registries varies. Detection of a change in frequency over time has led to specific epidemiological studies of birth defects (e.g., the Chernobyl accident, trisomy 21 in West Berlin, and neural tube defects in a small series in Turkey) that have not been replicated by other studies (Little, 1993).

There are also several cancer registries as part of the Surveillance, Epidemiology, and End Results Program (NCI, 2001). A health monitoring system that links such registries to exposures of interest may enhance the ability to detect unsuspected outcomes. Longitudinal data on food consumption patterns can at times be linked with health data via the use of health plan databases and hospital admissions and discharge records. The observed and expected rates of birth defects or specific diagnoses can be compared. It often is noted that these data are subject to potential biases, such as uncontrolled confounding and ecological fallacy. However, such analyses, if well designed, can use more sophisticated statistical techniques that allow inferences to be drawn.

Several authors have advocated the use of nonclassical model-based methods in this regard, including Bayesian methods, with applications to specific investigation of trends in breast cancer mortality (Bernardinelli et al.,1995), chronic myeloid leukemia (Chen, 1999), cancer of the oral cavity (Knorr-Held and Rasser,

2000), and particulate matter air pollution and premature mortality (Dominici et al., 2003).

There are no routine systems of cardiovascular morbidity ascertainment, although there are several large epidemiological cohort studies that include the endpoints of stroke or myocardial infarction, such as the Framingham study, the Nurses Health Study, the Women's Health Initiative, the Physicians Health Study, and the Cardiovascular Health Study (Abbott et al., 1988; Ma et al., 1999; Psaty et al., 1999; Rimm et al., 1998; Rossouw et al., 2002). Mortality data are collected routinely at the local level and aggregated into county and state statistics with associated age, gender, and other sociodemographic information. By extracting information from national surveys such as NHANES, it is possible to impute values for functional status or quality of life, such as the Health Utility Index, or the SF36. These in turn can be integrated with mortality information to calculate quality-adjusted life years.

DISCUSSION

The ability to evaluate the unintended health effects of a genetically engineered organism that expresses a significantly different phenotype than its conventional counterpart is problematic (Kuiper et al., 2001). As recognized by other expert panels, current risk-assessment paradigms and drug-safety evaluation programs are inappropriate methods to apply to the determination of the potential for unintended adverse health effects of GE food (Atherton, 2002; FAO/WHO, 2000).

Unlike chemicals and drugs, a dose-response relationship cannot be established for food (i.e., it is not possible to dose an animal with 10, 1, and 0.1 times the volume of food). Foods also represent complex mixtures that must be tested as a whole to consider possible nonadditive interactions that can significantly impact toxicological outcome (Dybing et al., 2002). Consequently, even though these technologies may satisfy the hazard identification step of risk evaluation, there are no existing validated methods for dose-response characterization for a complex mixture such as food.

New approaches should be based on a risk-assessment strategy proposed by the National Research Council (NRC, 1983) and rely on "substantial equivalence" to illustrate distinctions that may exist between foods modified by genetic engineering compared with those modified through traditional (non-GE) methods. Further, such evaluations would be expected at all stages of product development, including gene discovery, selection, and advancement to commercialization, and followed by postmarketing studies to further assess both intended and unintended effects. Epidemiological studies may be helpful in the postmarketing phase, provided they are conducted with the rigor that contemporary methods allow. The more definition that can be brought to bear with respect to defining exposure, the greater the inferential potential of these observational population-based methods.

Several new statistical methods are under development in an effort to address the need for tools to use in pursuit of functional genomics in a systematic and rigorous fashion in order to understand better the complex pathways of basic physiology and disease etiology (Boguski and McIntosh, 2003). These new techniques are expected to have applicability to the types of postmarketing studies presented in this chapter.

REFERENCES

Abbott RD, Wilson PW, Kannel WB, Castelli WP. 1988. High density lipoprotein cholesterol, total cholesterol screening, and myocardial infarction. The Framingham Study. *Arteriosclerosis* 8:207–211.

AgBiotechNet. 2000. *"Golden Mustard" to Combat Vitamin A Deficiency.* Online. Available at http://www.agbiotechnet.com/news/archive/2000/general/1512.asp#plant2. Accessed April 16, 2003.

Alberts DS, Martinez ME, Roe DJ, Guillen-Rodriquez JM, Marshall JR, van Leuuwen JB, Reid ME, Ritenbaugh C, Vargas PA, Bhattacharyya AB, Earnest DL, Sampliner RE. 2000. Lack of effect of a high-fiber cereal supplement on the recurrence of colorectal adenomas. Phoenix Colon Cancer Prevention Physicians' Network. *N Engl J Med* 342:1156–1162.

Anderson JJ, Anthony MS, Cline JM, Washburn SA, Garner SC. 1999. Health potential of soy isoflavones for menopausal women. *Public Health Nutr* 2:489–504.

Anderson JT, Grande F, Keys A. 1961. Hydrogenated fats in the diet and lipids in the serum of man. *J Nutr* 75:388–394.

Anderson LM, Diwan BA, Fear NT, Roman E. 2000. Critical windows of exposure for children's health: Cancer in human epidemiological studies and neoplasms in experimental animal models. *Environ Health Perspect* 108S:573–594.

APHIS (Animal and Plant Health Inspection Service). 1998. *Determination of Nonregulated Status for Bt Cry9C Insect Resistant and Glufosinate Tolerant Corn Transformation Event CBH-351.* Online. U.S. Department of Agriculture. Available at www.aphis.usda.gov/brs/dec_docs/9726501p_det.htm. Accessed December 17, 2002.

Aro A, Jauhiainen M, Partanen R, Salminen I, Mutanen M. 1997. Stearic acid, trans fatty acids, and dairy fat: Effects on serum and lipoprotein lipids, apolipoproteins, lipoprotein (a), and lipid transfer proteins in healthy subjects. *Am J Clin Nutr* 65:1419–1426.

Ascherio A, Rimm EB, Giovannucci EL, Spiegelman D, Stampfer MJ, Willett WC. 1996. Dietary fat and risk of coronary heart disease in men: Cohort follow-up study in the United States. *Br Med J* 313:84–90.

Atherton KT. 2002. Safety assessment of genetically modified crops. *Toxicology* 181–182:421–426.

Beever DE, Kemp CF. 2000. Safety issues associated with the DNA in animal feed derived from genetically modified crops. A review of scientific and regulatory procedures. *Nutr Abstr Rev Series B: Livestock Feeds and Feeding* 70:175-182.

Beier RC. 1990. Natural pesticides and bioactive components in foods. *Rev Environ Contam Toxicol* 113:47–137.

Bernardinelli L, Montomoli C, Ghislandi M, Pascutto C. 1995. Bayesian analysis of ecological studies. *Epidemiol Prev* 19:175–189.

Bertazzi PA, Consonni D, Bachetti S, Rubagotti M, Baccarelli A, Zocchetti C, Peatori AC. 2001. Health effects of dioxin exposure: A 20-year mortality study. *Am J Epidemiol* 153:1031–1044.

Beyer P, Al-Babili S, Ye X, Lucca P, Schaub P, Welsch R, Potrykus I. 2002. Golden rice: Introducing the beta-carotene biosynthesis pathway into rice endosperm by genetic engineering to defeat vitamin A deficiency. *J Nutr* 132:506S–510S.

Bingham SA. 1991. Limitations of the various methods for collecting dietary intake data. *Ann Nutr Metab* 35:117–127.

Bingham SA. 2002. Biomarkers in nutritional epidemiology. *Public Health Nutr* 5:821–827.

Bingham SA, Day NE, Luben R, Ferrari P, Slimani N, Norat T, Clavel-Chapelon F, Kesse E, Nieters A, Boeing H, Tjonneland A, Overvad K, Martinez C, Dorronsoro M, Gonzalez CA, Key TJ, Trichopoulou A, Naska A, Vineis P, Tumino R, Krogh V, Bueno-de-Mesquita HB, Peeters PHM, Berglund G, Hallmans G, Lund E, Skeie G, Kaaks R, Riboli E. 2003. Dietary fibre in food and protection against colorectal cancer in the European Prospective Investigation into Cancer and Nutrition (EPIC): An observational study. *Lancet* 361:1496–1501.

Biosafety Clearinghouse. 2003. Belgian Biosafety Server. Online. Available at http://biosafety.ihe.be/ARGMO/Documents/bijlage8.pdf. Accessed September 15, 2003.

Boguski MS, McIntosh MW. 2003. Biomedical informatics for proteomics. *Nature* 422:233–237.

Breiteneder H, Ebner C. 2000. Molecular and biochemical classification of plant-derived food allergens. *J Allergy Clin Immunol* 106:27–36.

Bucchini L, Goldman LR. 2002. Starlink corn: A risk analysis. *Environ Health Perspect* 110:5–13.

Burks AW, Fuchs RL. 1995. Assessment of the endogenous allergens in glyphosate-tolerant and commercial soybean varieties. *J Allergy Clin Immunol* 96:1008–1010.

Burks AW, Williams LW, Helm RM, Connaughton C, Cockrell G, O'Brien T. 1991. Identification of a major peanut allergen, Ara h 1, in patients with atopic dermatitis and positive peanut challenges. *J Allergy Clin Immunol* 88:172–179.

Butchko HH, Tschanz C, Kotsonis FN. 1994. Postmarketing surveillance of food additives. *Regul Toxicol Pharmacol* 20:105–118.

Chen R. 1999. The cumulative *q* interval curve as a starting point in disease cluster investigation. *Stat Med* 18:3299–3307.

Codex Alimentarius Commission. 2002. *Report of the Third Session of the Codex Ad Hoc Intergovernmental Task Force on Foods Derived from Biotechnology.* ALINORM 03/34. Rome: Food and Agriculture Organization of the United Nations.

Copple B. 2002. Monsanto has finally found some big markets abroad for its biotech crops. *Forbes* 170:48.

Dearman RJ, Caddick H, Basketter DA, Kimber I. 2000. Divergent antibody isotype responses induced in mice by systemic exposure to proteins: A comparison of ovalbumin with bovine serum albumin. *Food Chem Toxicol* 38:351–360.

Diplock AT, Aggett PJ, Ashwell M, Bornet F, Fern EB, Roberfroid MB. 1999. Scientific concepts of functional foods in Europe: Consensus document. *Br J Nutr* 81:S1–S27.

Doerfler W, and Schubbert R. 1997. Fremde DNA im Saugersystem. *Deut Arzt* 94:51-52.

Dominici F, McDermott A, Zeger SL, Samet JM. 2003. Airborne particulate matter and mortality: Timescale effects in four US cities. *Am J Epidemiol* 157:1055–1065.

Dribneki JCP, Green AG, Atlin GN. 1996. Linola 989 low linolenic flax. *Can J Plant Sci* 76:329–331.

Drinkard CR, Shatin D, Clouse J. 2000. Postmarketing surveillance of medications and pregnancy outcomes: Clarithromycin and birth malformations. *Pharmacoepidemiol Drug Saf* 9Z:549–556.

Dybing E, Doe J, Groten J, Kleiner J, O'Brien J, Renwick AG, Schlatter J, Steinberg P, Tritscher A, Walker R, Younes M. 2002. Hazard characterisation of chemicals in food and diet: Dose response, mechanisms and extrapolation issues. *Food Chem Toxicol* 40:237–282.

Efron B, Tibshirani R. 2002. Empirical Bayes methods and false discovery rates for microarrays. *Genet Epidemiol* 23:70–86.

Ermel RW, Kock M, Griffey SM, Reinhart GA, Frick OL. 1997. The atopic dog: A model for food allergy. *Lab Anim Sci* 47:40–49.

Falco SC, Guida T, Locke M, Mauvais J, Sanders C, Ward RT, Webber P. 1995. Transgenic canola and soybean seeds with increased lysine. *Biotechnology* 13:577–582.

FAO/WHO (Food and Agriculture Organization of the United Nations/World Health Organization). 1996. *Biotechnology and Food Safety. Report of a Joint FAO/WHO Consultation.* Rome: FAO.

FAO/WHO. 2000. *Safety Aspects of Genetically Modified Foods of Plant Origin. Report of a Joint FAO/WHO Expert Consultation on Foods Derived from Biotechnology.* Geneva: WHO.

FAO/WHO. 2001. *Evaluation of the Allergenicity of Genetically Modified Foods. Report of a Joint FAO/WHO Expert Consultation.* Rome: FAO.

FAO/WHO. 2003. Report of the Fourth Session of the Codex *Ad Hoc* Intergovernmental Task Force on Foods Derived from Biotechnology, Yokohama, Japan. Online. Available at ftp://ftp.fao.org/docrep/fao/meeting/006/y9220e.pdf. Accessed September 21, 2003.

FDA (Food and Drug Administration). 1996. Food additives permitted for direct addition to food for human consumption: Olestra; Final rule. *Fed Regis* 62:3118–3173.

Federal Register. 1992. Statement of Policy: Foods Derived from New Plant Varieties. 57 FR 22984; (21 U.S.C. 342(a) (1)); and 21 (U.S.C 321(s)). Washington, DC: Government Printing Office.

Fuchs RL, Ream JE, Hammond BG, Naylor MW, Leimgruber RM, Berberich SA. 1993. Safety assessment of the neomycin phosphotransferase II (NPT II) protein. *Biotechnology (N Y)* 11:1543–1547.

GISSI Trial Group. 1999. Dietary supplementation with *n*-3 polyunsaturated fatty acids and vitamin E after myocardial infarction: Results of the GISSI-Prevenzione Trial. *Lancet* 354:447–455.

Goldwyn S, Lazinsky A, Wei H. 2000. Promotion of health by soy isoflavones: Efficacy, benefit and safety concerns. *Drug Metabol Drug Interact* 17:261–289.

Green A. 1994. Linola. *Lipid Technol* 6:29–33.

Harrison LA, Bailey MR, Naylor MW, Ream JE, Hammond BG, Nida DL, Burnette BL, Nickson TE, Mitsky TA, Taylor MA, Fuchs RL, Padgette SR. 1996. The expressed protein in glyphosate-tolerant soybean, 5-enolpyruvylshikimate-3-phosphate synthase from *Agrobacterium* sp. strain CP4, is rapidly digested in vitro and is not toxic to acutely gavaged mice. *J Nutr* 126:728–740.

Helm RM. 2002. Food allergy animal models: An overview. *Ann N Y Acad Sci* 964:139–150.

Hitz WD, Carlson TJ, Kerr PS, Sevastian SA. 2002. Biochemical and molecular characterization of a mutation that confers a decreased raffinosaccharide and phytic acid phenotype on soybean seeds. *Plant Physiol* 128:650–660.

Hochberg Y. 1988. A sharper Bonferroni procedure for multiple tests of significance. *Biometrika* 75:800–802.

Holcroft CA, Spiegelman D. 1999. Design of validation studies for estimating the odds ratio of exposure-disease relationships when exposure is misclassified. *Biometrics* 55:1193–1201.

Hu F, Willett WC. 2002. Optimal diets for prevention of coronary heart disease. *J Am Med Assoc* 288:2569–2578.

IENICA (Interactive European Network for Industrial Crops and Their Applications). 2003. Linola: Edible Linseed. Online. Available at http://www.ienica.net/crops/linola.htm. Accessed August 26, 2003.

IFT (Institute of Food Technologists). 2000. IFT expert report on biotechnology and foods. Human food safety of rDNA biotechnology-derived foods. *Food Technol* 54:53–61.

IOM (Institute of Medicine). 1998. *Dietary Reference Intakes: A Risk Assessment Model for Establishing Upper Intake Levels for Nutrients.* Washington, DC: National Academy Press.

IOM. 2001. *Dietary Reference Intakes: Applications in Dietary Assessment.* Washington, DC: National Academy Press.

IOM. 2003. *Dietary Reference Intakes: Applications in Dietary Planning.* Washington, DC: The National Academies Press.

Ito K, Inagaki-Ohara K, Murosaki S, Nishimura H, Shimokata T, Torii S, Matsuda T, Yoshikai Y. 1997. Murine model of IgE production with a predominant Th2-response by feeding protein antigen without adjuvants. *Eur J Immunol* 27:3427–3437.

Judd JT, Baer DJ, Clevidence BA, Muesing RA, Chen SC, Westrate JA, Meijer GW, Wittes J, Lichenstein AH, Vilellabach M, Schaefer EJ. 1998. Effects of margarine compared with those of butter on blood lipid profiles related to cardiovascular disease risk factors in normolipemic adults fed controlled diets. *Am J Clin Nutr* 68:768–777.

Kaput J, Rodriguez RL. 2004. Nutritional genomics: The next frontier in the postgenomic era. *Physiol Genomics* 16:166–177.

Katan MB. 2000. Trans fatty acids and plasma lipoproteins. *Nutr Rev* 58:188–191.

Kelsey JL. 1996. *Methods in Observational Epidemiology*. 2nd ed. New York: Oxford University Press.

Keys A, Brozek J, Henschel A, Mickelson O, Taylor HL. 1950. *The Biology of Human Starvation*. Minneapolis: University of Minnesota Press.

Kimber I, Dearman RJ, Penninks AH, Knippels LMJ, Buchanan BB, Hammerberg B, Jackson HA, Helm RM. 2003. Assessment of protein allergenicity based on immune reactivity: Animal models. *Environ Health Perspect* 111:1125–1130.

Kinney AJ. 1996. Development of genetically engineered soybean oils for food applications. *J Food Lipids* 3:273–292.

Kipnis V, Midthune D, Freedman L, Bingham S, Day NE, Riboli E, Ferrari P, Carroll RJ. 2002. Bias in dietary-report instruments and its implications for nutritional epidemiology. *Public Health Nutr* 5:915–923.

Klaasen CD, Amdur MO, Doull J. 1986. *Casarett and Doull's Toxicology: The Basic Science of Poisons*. 3rd ed. New York: Macmillan.

Knippels LM, Penninks AH. 2003. Assessment of the allergic potential of food protein extracts and proteins on oral application using the brown Norway rat model. *Environ Health Perspect* 111:233–238.

Knippels LM, Penninks AH, van Meeteren M, Houben GF. 1999. Humoral and cellular immune responses in different rat strains on oral exposure to ovalbumin. *Food Chem Toxicol* 37:881–888.

Knorr-Held L, Rasser G. 2000. Bayesian analysis of ecological studies. *Biometrics* 56:13–21.

Kok EJ, Kuiper HA. 2003. Comparative safety assessment for biotech crops. *Trends Biotechnol* 21:439–443.

Kristal AR, Patterson RE, Neuhouser ML, Thornquist MT, Neumark-Sztainer D, Rock CL, Berlin MC, Cheskin L, Schreiner PJ. 1998. Olestra Postmarketing Surveillance Study: Design and baseline results from the sentinel site. *J Am Diet Assoc* 98:1290–1296.

Kromhout D, Menotti A, Bloemberg B, Aravani C, Blackburn H, Buzina R, Dontas AS, Fidanza F, Giampaoli AS, Fidanzq F, Jansen AJ. 1995. Dietary saturated and trans fatty acids and cholesterol and 25-year mortality from coronary heart disease: The Seven Countries Study. *Prev Med* 24:308–315.

Kuiper HA, Kleter GA, Noteborn HP, Kok EJ. 2001. Assessment of the food safety issues related to genetically modified foods. *Plant J* 27:503–528.

Leon AS, Hunninghake DB, Bell C, Rassin DK, Tephly TR. 1989. Safety of long-term large doses of aspartame. *Arch Intern Med* 149:2318–2324.

Li XM, Schofield B, Huang CK, Kleiner GI, Sampson HA. 1999. A murine model of IgE-mediated cow's milk hypersensitivity. *J Allergy Clin Immunol* 103:206–214.

Lilienfeld DE. 1994. *Foundations of Epidemiology*. 3rd ed. New York: Oxford University Press.

Little J. 1993. The Chernobyl accident, congenital anomalies and other reproductive outcomes. *Paediatr Perinat Epidemiol* 7:121–151.

LSRO (Life Sciences Research Office). 1995. *Third Report on Nutrition Monitoring in the United States*. Vol 1. Washington DC: LSRO.

Ma J, Hennekens CH, Ridker PM, Stampfer MJ. 1999. A prospective study of fibrinogen and risk of myocardial infarction in the Physicians' Health Study. *J Am Coll Cardiol* 33:1347–1352.

McClintock JT, Schaffer CT, Sjoblad RD. 1995. A comparative review of the mammalian toxicity of *Bacillus thuringiensis*-based pesticides. *Pestic Sci* 45:95–105.

Mensink RP, Katan MB. 1990. Effect of trans fatty acids on high-density and low-density lipoprotein cholesterol levels in healthy subjects. *N Engl J Med* 323:439–445.

Metcalfe DD, Astwood JD, Townsend R, Sampson HA, Taylor SL, Fuchs RL. 1996. Assessment of the allergenic potential of foods derived from GE crop plants. *Crit Rev Food Sci Nutr* 36S:165–186.

MHLW (Ministry of Health, Labor, and Welfare of the Japanese Government). 2000. *Mandatory Requirement for Safety Assessment of Foods and Food Additives Produced by Recombinant DNA Techniques: Procedure of Application for Safety Assessment on Foods and Food Additives Produced by Recombinant DNA Techniques.* Announcement No.233, 1 May 2000. Online. Available at http://www.mhlw.go.jp/english/topics/food/sec03.html. Accessed August 23, 2003..

NCI (National Cancer Institute). 2001. Surveillance, Epidemiology, and End Results. Bethesda, MD: National Cancer Institute, National Institutes of Health. Online. Available at http://seer.cancer.gov/. Accessed December 12, 2003.

Nestle M. 2003. *Safe Food: Bacteria, Biotechnology, and Bioterrorism.* Berkeley: University of California Press.

Nielsen KM, Bones AM, Smalla K, van Elsas JD. 1998. Horizontal gene transfer from transgenic plants to terrestrial bacteria—a rare event? *FEMS Microbiol Rev* 22:79–103.

Nordlee JA, Taylor SL, Townsen, JA, Thomas LA, Bush RK. 1996. Identification of Brazil nut allergen in transgenic soybeans. *N Engl J Med* 334:688–692.

Noteborn HP, Kuiper HA. 1994. Safety assessment strategies for genetically-modified plant products. Case study: *Bacillus thuringiensis*-toxin tomato. In: Jones D, ed. *Proceedings of the 3rd International Symposium on the Biosafety Results of Field Tests of Genetically Modified Plants and Microorganisms.* Oakland, CA: Division of Agriculture and Natural Resources, University of California. Pp. 199–207.

NRC (National Research Council). 1983. *Risk Assessment in the Federal Government: Managing the Process.* Washington, DC: National Academy Press.

NRC. 1994. *Science and Judgment in Risk Assessment.* Washington, DC: National Academy Press.

OECD (Organization for Economic Cooperation and Development). 1993. *Safety Evaluation of Foods Derived by Modern Biotechnology—Concepts and Principles.* Paris: OECD.

OECD. 2000. *Report of the Task Force for the Safety of Novel Foods and Feeds.* Paris: OECD.

OECD. 2001. *Consensus Document on Compositional Considerations for New Varieties of Soybean: Key Food and Feed Nutrients and Anti-Nutrients.* Online. Available at www.olis.oecd.org/olis/2001doc.nsf/43bb6130e5e86e5fc12569fa005d004c/cdb400c627da47a5c1256b17002f840d/$FILE/JT00117705.PDF. Accessed January 16, 2003.

OFAS (Office of Food Additive Safety). 2001. *Toxicological Principles for the Safety Assessment of Direct Food Additives and Color Additives Used in Food. Redbook II-Draft.* Washington, DC: OFAS, Center for Food Safety and Applied Nutrition, Food and Drug Administration.

Oomen A, Ocke MC, Feskens JM, van Erp-Barrt MJ, Kok FJ, Dromhout D. 2001. Association between *trans* fatty acid intake and 10-year risk of coronary heart disease in the Zutphen Elderly Study: A prospective population-based study. *Lancet* 357:746–751.

OPPTS (Office of Prevention, Pesticides, and Toxic Substances). 1996. *OPPTS Harmonized Test Guidelines. Series 870. Health Effects Test Guidelines.* Washington, DC: U.S. Environmental Protection Agency.

Parent ME, Siemiatycki J, Fritschi L. 2000. Workplace exposures and oesophageal cancer. *Occup Environ Med* 57:325–334.

Pennie WD. 2002. Custom cDNA microarrays; Technologies and applications. *Toxicology* 181–182:551–554.

Peters U, Sinha R, Chatterjee N, Subar AF, Ziegler RG, Kulldorff M, Bresalier R, Weissfeld JL, Flood A, Schatzkin A, Hayes RB. 2003. Dietary fibre and colorectal adenoma in a colorectal cancer early detection programme. *Lancet* 361:1491–1495.

Pietinen P, Ascherio A, Korhonen P. 1997. Intake of fatty acids and risk of coronary heart disease in a cohort of Finnish men: The ATBC study. *Am J Epidemiol* 145:876–887.

Psaty BM, Furberg CD, Kuller LH, Bild DE, Rautaharju PM, Polak JF, Bovill E, Gottdiener JS. 1999. Traditional risk factors and subclinical disease measures as predictors of first myocardial infarction in older adults: The Cardiovascular Health Study. *Arch Intern Med* 159:1339–1347.

Rimm EB, Stampfer MJ, Ascherio A, Giovannucci E, Colditz GA, Willett WC. 1993. Vitamin E consumption and the risk of coronary heart disease in men. *N Engl J Med* 328:1450–1456.

Rimm EB, Willett WC, Hu FB, Sampson L, Colditz GA, Manson JE, Hennekens C, Stampfer MJ. 1998. Folate and vitamin B_6 from diet and supplements in relation to risk of coronary heart disease among women. *J Am Med Assoc* 279:359–364.

Rossouw JE, Anderson GL, Prentice RL, LaCroix AZ, Kooperberg C, Stefanick ML, Jackson RD, Beresford SA, Howard BV, Johnson KC, Kotchen JM, Ockene J. 2002. Risks and benefits of estrogen plus progestin in healthy postmenopausal women: Principal results from the Women's Health Initiative randomized controlled trial. *J Am Med Assoc* 288:321–333.

Rothman KJ, S Greenland, eds. 1998. *Modern Epidemiology*. 2nd ed. Philadelphia: Lippincott-Raven.

Rowan AJ, Shaywitz BA, Tuchman L, French JA, Luciano D, Sullivan CM. 1995. Aspartame and seizure susceptibility: Results of a clinical study in reportedly sensitive individuals. *Epilepsia* 36:270–275.

Rulis AM. 1992. Threshold of regulation—Options for handling minimal risk situations. In: Finley JW, Robinson SF, Armstrong DJ, eds. *Food Safety Assessment*. Washington, DC: American Chemical Society. Pp. 132–139.

Saks FM, Svetkey LP, Vollmer WM, Appel LJ, Bray GA, Harsha D, Obarzanek E, Conlin PR, Miller ER III, Simons-Morton DG, Karanaja N, Lin P. 2001. Effects on blood pressure of reduced dietary sodium and the Dietary Approaches to Stop Hypertension (DASH) diet. *N Engl J Med* 344:3–10.

Schatzkin A, Gail M. 2002. The promise and peril of surrogate end points in cancer research. *Nat Rev Cancer* 2:1–9.

Schatzkin A, Lanza E, Corle D, Lance P, Iber F, Caan B, Shike M, Weissfeld J, Burt R, Cooper MR, Kikendall JW, Cahill J. 2000. Lack of effect of a low-fat, high-fiber diet on the recurrence of colorectal adenomas. *N Engl J Med* 342:1149–1155.

Schilter B, Holzhauser D, Cavin C, Huggett AC. 1996. An integrated in vivo/in vitro strategy to improve food safety evaluation. *Trends Food Sci Technol* 7:327–332.

Schnepf E, Crickmore N, Van Rie J, Lereclus D, Baum J, Feitelson J, Zeigler DR, Dean DH. 1998. *Bacillus thuringiensis* and its pesticidal crystal proteins. *Microbiol Molec Biol Rev* 62:775–806.

Schubbert R, Hohlweg, U, Renz D, Doerfler W. 1998. On the fate of orally ingested foreign DNA in mice: Chromosomal association and placental transfer to the fetus. *Mol Gen Genet* 259:569–576.

Sissell K. 2002. Seed smugglers sow dissent in Brazil. *Chem Week* 164:21.

Slough CL, Niday RK, Zorich NL, Jones JK. 2001. Postmarketing surveillance of new food ingredients: Design and implementation of the program for the fat replacer Olestra. *Regul Toxicol Pharmacol* 33:218–223.

Smith T. 2003. Brazil to lift ban on crops with genetic modification. *N Y Times* 153:W1.

Spiegelman D, Carroll RJ, Kipnis V. 2001. Efficient regression calibration for logistic regression in main study/internal validation study designs with an imperfect reference instrument. *Stat Med* 20:139–160.

Stampfer MJ, Hennekens CH, Manson JE, Colditz GA, Rosner B, Willettt WC. 1993. Vitamin E consumption and the risk of coronary disease in women. *N Engl J Med* 328:1444–1449.

Stanley JS, King N, Burks AW, Huang SK, Sampson HA, Cockrell G, Helm RM, West CM, Bannon GA. 1997. Identification and mutational analysis of the immunodominant IgE binding epitopes of the major peanut allergen, Ara h 2. *Arch Biochem Biophys* 342:244–253.

Story JA, Savaiano DA. 2001. Dietary fiber and colorectal cancer: What is appropriate advice? *Nutr Rev* 59:84–86.

Stover PJ. 2004. Nutritional genomics. *Physiol Genomics* 16:161–165.

Svetkey LP, Simons-Morton D, Vollmer WM, Appel LJ, Conlin PR, Ryan DH, Ard J, Kennedy BM. 1999. Effects of dietary patterns on blood pressure: Subgroup analysis of the Dietary Approaches to Stop Hypertension (DASH) randomized clinical trial. *Arch Intern Med* 159:285–293.

Taylor SL. 2001. Safety assessment of genetically modified foods. *J Nematol* 33:178–182.

Taylor SL. 2002. Protein allergenicity assessment of foods produced through agricultural biotechnology. *Annu Rev Pharmacol Toxicol* 42:99–112.

Taylor SL, Hefle SL. 2002. Food allergy assessment for products derived through plant biotechnology. In: Thomas JA, Fuchs RL, eds. *Biotechnology and Safety Assessment.* 3rd ed. San Diego: Academic Press. Pp. 325–345.

Thomas JG, Olson JM, Tapscott SJ, Zhao LP. 2001. An efficient and robust statistical modeling approach to discover differentially expressed genes using genomic expression profiles. *Genome Res* 11:1227–1236.

Thompson FE, Byers Y. 1994. Dietary assessment resource manual. *J Nutr* 124:2245S–2317S.

Thornquist MD, Kristal AR, Patterson RE, Neuhuser ML, Rock CL, Neumark-Sztainer D, Cheskin LJ. 2000. Olestra consumption does not predict serum concentrations of carotenoids and fat-soluble vitamins in free-living humans: Early results from the sentinel site of the Olestra post-marketing surveillance study. *J Nutr* 130:1711–1718.

Tollefson L, Barnard RJ. 1992. An analysis of FDA passive surveillance reports of seizures associated with consumption of aspartame. *J Am Diet Assoc* 92:598–601.

Vedavanam K, Srijayanta S, O'Reilly J, Raman A, Wiseman H. 1999. Antioxidant action and potential antidiabetic properties of an isoflavonoid-containing soyabean phytochemical extract (SPE). *Phytother Res* 13:601–608.

Vergriese AJ. 1972. Dietary fat and cardiovascular disease: Possible modes of action of linoleic acid. *Proc Nutr Soc* 31:323–329.

Vesselinovitch SD, Rao KV, Mihailovich N. 1979. Neoplastic response of mouse tissues during perinatal age periods and its significance in chemical carcinogenesis. *J Natl Cancer Inst Monogr* 51:239–250.

Westrate JA, van het Hof KH. 1995. Sucrose polyester and plasma carotenoid concentrations in healthy subjects. *Am J Clin Nutr* 62:591-597.

WHO (World Health Organization). 1993. Health aspects of marker genes in genetically modified plants. Report of a WHO Workshop, Copenhagen, Denmark, September 21-24. Geneva, Switzerland: World Health Organization.

WHO. 1996. *Trace Elements in Human Nutrition and Health.* Geneva: WHO.

WHO. 2000. Release of genetically modified organisms in the environment: Is it a health hazard? Report of a Joint WHO/EURO-ANPA Seminar, September 7-9. Online. Geneva, Switzerland: World Health Organization. Available online at http://www.who.int/foodsafety/publications/biotech/en/ec_sept2000.pdf. Accessed May 17, 2003.

Willett W. 1998. *Nutritional Epidemiology.* 2nd ed. Oxford: Oxford University Press.

Willett WC, Ascherio A. 1994. Trans fatty acids: Are the effects only marginal? *Am J Public Health* 84:722–724.

Xu XL, Olson JM, Zhao LP. 2002. A regression-based method to identify differentially expressed genes in microarray time course studies and its application to an inducible Huntington's disease transgenic model. *Hum Mol Genet* 11:1977–1985.

Ye X, Al-Babili S, Kloti A, Zhang J, Lucca P, Beyer P, Potrykus I. 2000. Engineering the provitamin A (ß-carotene) biosynthetic pathway into (carotenoid-free) rice endosperm. *Science* 287:303–305.

Yusuf S, Dagnenais G, Pogue J, Bosch J, Sleight P. 2000. Vitamin E supplementation and cardiovascular events in high-risk patients. *N Engl J Med* 342:154–160.

Zitnak A, Johnston GR. 1970. Glycoalkaloid content of B5141-6 potatoes. *Am Potato J* 47:256–260.

7

Framework, Findings, and Recommendations

BACKGROUND

Genetic engineering is one of many genetic modification techniques that can be used to generate foods of altered composition, including novel components. The application of recombinant DNA technology allows a unique opportunity to introduce new genes into plants and animals used for food. However, the products of this technology are not always distinguishable from other methods of genetic modification. Moreover, application of any technique to produce altered levels of or novel food components can result in unintended compositional changes that may in turn result in an adverse health effect.

The safety assessment methods recommended in this chapter for genetically modified (GM) foods—including those that are genetically engineered (GE)— are intended to identify products with a greater likelihood for the potential to introduce unintended adverse health effects. Additional findings and recommendations specifically related to animal genetic manipulation and cloning are contained in the committee's subreport.

The following framework forms the basis of the committee's recommendations presented later in this chapter, as well as those presented in the subreport.

FRAMEWORK FOR ASSESSING
POTENTIAL UNINTENDED EFFECTS

Any GM food has the potential for producing levels of primary and secondary metabolites that differ from its parental counterparts. As part of its task, the committee developed a framework, illustrated by the flow chart in Figure 7-1, as a guide for considering appropriate questions and methods to determine potential

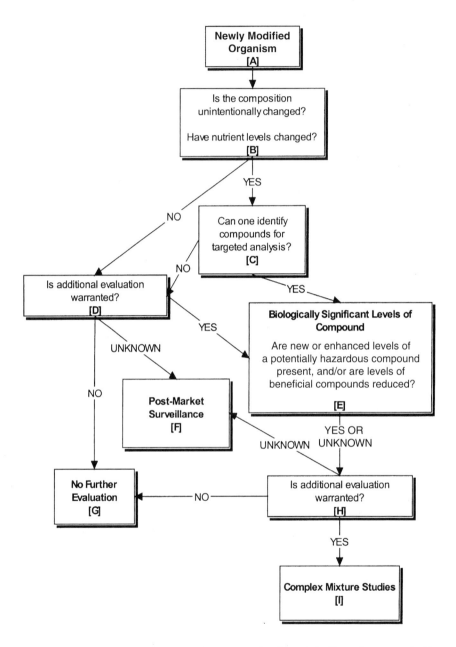

FIGURE 7-1   Flowchart for determining potential unintended effects from genetically modified foods.

unintended changes in the levels of endogenous nutrients, toxicants, allergens, or other compounds in all types of GM food—including GE food—that may lead to an unintended adverse health effect. It is important to note that this framework does not treat genetic engineering as a technology that is completely separate from other genetic modification techniques; the flow chart can usefully be applied to the full range of genetic modification technologies.

However, there are limitations to the application of this framework—or any other—because technological advances in analytical chemistry have exceeded our ability to interpret the consequences to human health of changes in food composition. Although compositional changes can be detected readily in food, and the power of profiling techniques is rapidly increasing our ability to identify compositional differences between GE food products and their conventional counterparts, methods for determining the biological relevance of these changes and predicting unintended adverse health effects are understudied. As discussed in this report, further advances in analytical technologies and their interpretation are needed to address these limitations.

Nevertheless, the committee believes that useful assessments currently can be made using this framework, giving consideration to questions such those listed below.

1. What differences exist from the progenitor line?

This question should address the known nutrients, toxicants, and antinutritional factors in order to identify and quantify changes introduced to food, whether intentional or unintentional. Two related questions exist:

    a. What is the relevant progenitor to use as a comparator?

    b. Should all detected differences trigger a requirement for further analytical work?

The selection of the relevant progenitor line is not a trivial issue. Because it is known that the average composition of food crops has changed over time as a result of breeding and changes in agricultural practices, selection of a historical progenitor is not appropriate. The immediate isogenic progenitor line is an appropriate comparator; however, the role of environmental factors on composition must be considered. Similarly, the progenitor's role in the total diet of target populations must be considered. It is the genome that enables environmental responses, and environmental variables often have been shown to have large effects on composition. The interaction of genotypic and environmental variables must be considered in evaluating compositional effects of genetic modification. Comparisons of the new line with the progenitor, when both are grown under a single set of environmental conditions, would be informative, but not conclusive.

It is proposed that compositional differences attributable to genetic changes be evaluated on a case-by-case basis. In particular, the importance of differences should be taken into consideration relative to the importance of the particular food as a source of a particular component. For example, the detection of a lower vitamin C content of a GE line of radishes should not be a cause for concern because radishes are not a significant source of vitamin C in human diets. In contrast, lower vitamin C in oranges or derived products, which are an important source of vitamin C in the U.S. diet, should trigger further consideration in that country.

2. What is the biological relevance of an identified compositional change from the perspective of human nutrition and health?

If differences between the progenitor and the new line are detected with respect to nutritionally or toxicologically significant constituents, these differences should be evaluated further to determine the importance of the differences relative to the compositional range and variability of the compound in major commercial varieties. A difference in composition from the progenitor that did not exceed the variability of major commercial varieties should not be considered a cause for concern.

3. What is the biological relevance of a compositional difference with respect to subgroups of the population who have either greater exposure or greater potential susceptibility to an unintended effect?

This question, in essence, constitutes a sensitivity analysis. The issues to be considered include:

a. Is the modified food a large component of average dietary intake?
b. Is the modified food a small component of average dietary intake?
c. What is the anticipated effect on upper-range (niche) consumers?

These questions on the interpretation of analytical data and related issues are discussed more fully in Chapters 5 and 6.

The following examples illustrate how newly modified organisms would proceed through the process presented in the flow chart, using hypothetical scenarios to illustrate the application of various approaches.

In the routine breeding process of crossing—for example, when one wheat variety is crossed with another to transfer a disease resistance gene—the resulting variety is genetically modified, but the desired trait that is obtained is not anticipated to be new to the species or to the food supply. This product would be considered in light of the questions posed in Boxes A through C. If the response to these questions is "no," and if no other novel substances of concern are present, the products flow from Box D to G; such a variety need not trigger additional concerns.

As an alternate example, a cross of the potatoes *Solanum tuberosum* and *S. brevidans* resulting in the production of demissidine—a novel toxic substance in the tubers—would warrant further scrutiny because the new potato variety is expressing a toxic substance that is new to both the species and the food supply. The identification of demissidine would require further testing (Boxes D and E), such as toxicological or complex mixture studies (Boxes H and I).

In another example, GE rice that expresses a soybean gene but carries no substance new to the species or to the food supply would proceed from Box A through Box C, as in the wheat example above. However, since soybeans are a common allergen, there would be need for additional scrutiny to determine the potential for allergenic response to the product. Once the potential for transfer of a soybean allergen into the rice is determined and, if appropriate, action is taken, the evaluation process would not go any further. Alternatively, a GE soybean that expresses a daffodil gene to enhance production of beta-carotene does warrant additional evaluation because the expression of beta-carotene, a precursor to an essential micronutrient, is substantially enhanced (Box E), even though it is not a novel substance in the food supply. In this case, the answer to the question posed in Box C could be either "yes" or "no," and if "no," the process would proceed to Box D. If the level of beta-carotene expressed by the soybean is at a level known to have biological significance, the evaluation process would proceed from Box D to Box E, then through Boxes H and I. If the biological significance is "unknown," the evaluation process would proceed from Box E to Box H to Box F (postmarket surveillance). Similarly, if unintended compositional changes are accompanied in the introduction of this novel gene to the soybean, the nature of the change could be evaluated within the suggested framework.

FINDINGS AND RECOMMENDATIONS

Overall Findings and Recommendation

Findings

All new crop varieties, animal breeds (see cloning subreport), and microbial strains carry modified deoxyribonucleic acid (DNA) that differs from parental strains. Methods to genetically modify plants, animals, and microbes are mechanistically diverse and include both natural and human-mediated activities. Health outcomes could be associated with the presence or absence of specific substances added or deleted using genetic modification techniques, including genetic engineering, and with unintended compositional changes.

The likelihood that an unintended compositional change will occur can be placed on a continuum that is based on the method of genetic modification used (see Figure 3-1). The genetic modification method used, however, should not be the sole criterion for suspecting and subsequently evaluating possible health effects associated with unintended compositional changes.

All evidence evaluated to date indicates that unexpected and unintended compositional changes arise with all forms of genetic modification, including genetic engineering. Whether such compositional changes result in unintended health effects is dependent upon the nature of the substances altered and the biological consequences of the compounds. To date, no adverse health effects attributed to genetic engineering have been documented in the human population.

Recommendation 1

The committee recommends that compositional changes that result from all genetic modification in food, including genetic engineering, undergo an appropriate safety assessment. The extent of an appropriate safety assessment should be determined prior to commercialization. It should be based on the presence of novel compounds or substantial changes in the levels of naturally occurring substances, such as nutrients that are above or below the normal range for that species (see Chapter 3), taking into account the organism modified and the nature of the introduced trait.

Safety Assessment Tools for Assessing Unintended Effects Prior to Commercialization

Findings

Current voluntary and mandated safety assessment approaches focus primarily on intended and predictable effects of novel components of GE foods. Introduction of novel components into food through genetic engineering can pose unique problems in the selection of suitable comparators for the analytical procedures that are crucial to the identification of unintended compositional changes. Other jurisdictions, particularly the European Union, evaluate all GE food products prior to commercialization, but exempt from similar evaluation all other GM foods. As previously discussed in Chapter 3, the policy to assess products based exclusively on their method of breeding is scientifically unjustified.

The most appropriate time for safety assessment of all new food is in the premarket period prior to commercialization, although verification of safety assessments may continue in the postmarket period, generally in cases when a potential problem has been identified or if there is elevated cause for concern. Examples of specific premarket assessments of newly introduced compositional changes to selected GE food are:

- protein, fat, carbohydrate, fiber, ash, and water in a proximate analysis;
- essential macro- and micronutrients in a nutritional analysis;
- known endogenous toxicants and antinutrients in specific species;
- endogenous allergens;
- other naturally occurring, species-specific constituents of potential inter-

est, such as isoflavones and phytoestrogens in soybean or alkaloids in tomato or potato;

- gross agronomic characteristics;
- data derived from domestic animal feeding trials to assess the nutritional quality of new crops; and
- data derived from toxicological studies in animals.

Recommendation 2

The committee recommends that the appropriate federal agencies determine if evaluation of new GM foods for potential adverse health effects from both intended and unintended compositional changes is warranted by elevated concern, such as identification of a novel substance or levels of a naturally occurring substance that exceeds the range of recommended or tolerable intake.

Recommendation 3

For those foods warranting further evaluation, the committee recommends that a safety assessment be conducted prior to commercialization and continued evaluation postmarket where safety concerns are present. Specifically, the committee recommends the following safety assessment actions.

- Develop a paradigm for identifying appropriate comparators for GE food.
- Collect and make publicly available key compositional information on essential nutrients, known toxicants, antinutrients, and allergens of commonly consumed varieties of food (see the Research Needs section, later in this chapter). These should include mean values and ranges that typically occur as a function of genetic makeup, differences in physiological state, and environmental variables.
- Remove compositional information on GE foods from proprietary domains to improve public accessibility.
- Continue appropriate safety assessments after commercialization to verify premarket evaluations, particularly if the novelty of the introduced substance or the level of a naturally occurring substance leads to increased safety concerns.

Analytical Methodologies

Findings

During the past decade, analytical methodologies for separating and quantifying messenger ribonucleic acids (mRNA), proteins, and metabolites have improved markedly. Applying these methodologies to the targeted analysis of known nutrients and toxicants will improve the knowledge base for these food constituents. The broad application of targeted methods and continuing development of profiling methods will provide extensive information about food composition and further improve the knowledge base of defined chemical food constituents. The knowledge and understanding needed to relate such compositional information to potential unintended health effects is far from complete, however. Furthermore, currently available bioinformatics and predictive tools are inadequate for correlating compositional analyses with biological effects.

Analytical profiling techniques are appropriate for establishing compositional differences among genotypes, but they must also take into account modification of the profile obtained due to genotype-by-environmental interactions (the influence of the environment on expression of a particular genotype). The knowledge base required to interpret results of profiling methods, however, is insufficiently developed to predict or directly assess potential health effects associated with unintended compositional changes of GM food, as is the necessary associative information (e.g., proteomics, metabolomics, and signaling networks). Additionally, predictive tools to identify the expected behavior of complex and compound structures are limited and require a priori knowledge of their chemical structure, their biological relevance, and their potential interactive targets.

Recommendation 4

The committee recommends the development and employment of standardized sampling methodologies, validation procedures, and performance-based techniques for targeted analyses and profiling of GM food performed in the manner outlined in the flow chart shown in Figure 7-1. Sampling methodology should include suitable comparisons to the near isogenic parental variety of a species, grown under a variety of environmental conditions, as well as ongoing assessment of commonly consumed commercial varieties of food. These include:

- Reevaluation of current methodologies used to detect and assess the biological consequences of unintended changes in GM food, including better tools for toxicity assessment and a more robust knowledge base for determining which novel or increased naturally occurring components of food have a health impact.
- Use of data collection programs, such as the Continuing Survey of Food Intakes by Individuals and the National Health and Nutrition Examination Survey (NHANES), to collect information, prior to commercial release of a new GM food, on current food and nutrient intakes and exposure to known toxins or toxicants through food consumption. The information collected should be used to identify food consumption patterns in the general population and susceptible population subgroups that indicate a potential for adverse reactions to novel substances or increased levels of naturally occurring compounds in GM food.

Additional Tools for Postcommercialization:
Identification and Assessment of Unintended Effects

Findings

Postcommercialization or postmarket evaluation tools for verifying and validating premarket assessments of novel substances in food or detectable changes in diet composition, including tracking and epidemiological studies, are important components of the overall assessment of food safety. These tools provide a way to check the efficacy of premarket compositional and safety evaluations through a feedback process. In addition, information databases that result from postmarket studies can be valuable assets in the development of future premarket safety assessment tools.

Postmarket surveillance is a commonly accepted procedure, for example, with new pharmaceuticals and has been beneficial in the identification of harmful and unexpected side effects. As a result, pharmacologists accept postmarket surveillance as a part of the process to identify unexpected adverse outcomes from their products. This example is especially pertinent to GE foods because of the unique ability of this process to introduce gene sequences to generate novel products into organisms intended for use as food and especially in situations where the novel products are introduced at levels that have the potential to alter dietary intake patterns (e.g., elevated levels of key nutrients).

Given the possibility that food with unintended changes may enter the marketplace despite premarket safety mechanisms, postmarket surveillance of exposures and effects is needed to validate premarket evaluations. On the other hand, there are many instances in which postmarket surveillance may not be warranted. For example, when compositional comparisons of a new GM crop or food (e.g., Roundup Ready soybeans) with its conventional counterpart indicate they are compositionally very similar, exposure to novel components remains very low. Thus the process of identifying unintended compositional changes in food is best served by combining premarket testing with postmarket surveillance, when compositional changes indicate that it is warranted, in a feedback loop that follows a new GM food or food product long-term, from development through utilization (see Figure 7-1).

Recommendation 5

When warranted by changes such as altered levels of naturally occurring components above those found in the product's unmodified counterpart, population-specific vulnerabilities, or unexplained clusters of adverse health effects, the committee recommends improving the tracking of potential health consequences from commercially available foods that are genetically modified, including those that are genetically engineered, by actions such as the following:

• Improve the ability to identify populations that are susceptible to food allergens and develop databases relevant to tracking the prevalence of food allergies and intolerances in the general population, and in susceptible population subgroups.

• Improve and include other postmarket resources for identifying and tracking unpredicted and unintended health effects from GM foods:

— Improve the sensitivity of surveys and other analytical methodologies currently used to detect consumer trends in the purchase and use of GM foods after release into the marketplace,

— Standardize methods for monitoring reports of allergenicity to new foods introduced into the marketplace and apply them to new GM foods,

— Assure that current food labeling includes relevant nutritional attributes so that consumers can receive more complete information about the nutritional components in GM foods introduced into the marketplace, and

— Improve utilization of potential traceability technology, such as bar coding of animal carcasses and other relevant foods.

• Develop a database of unique genetic sequences (DNA, polymerase chain reaction sequences) from GE foods entering the marketplace to enable their identification in postmarket surveillance activities.

• Utilize existing nationwide food intake and health assessment surveys, including NHANES, to:

— Collect comparative information on diet and consumption patterns of the general population and ethnic subgroups in order to account for anthropological differences among population groups and geographic areas where GM foods may be consumed in skewed quantities, recognizing that this will be possible only under selected circumstances where intakes are not evenly distributed across population subgroups of interest and the relevant outcome data are available, and

— Provide better representation of the long-term nutritional and other health status information on a full range of children and ethnic groups whose intakes may differ significantly from those of the general population to determine whether changes in health status have occurred as a consequence of consuming novel substances or increased levels of naturally occurring compounds in GM foods released into the marketplace, recognizing again that this will be possible only under selected circumstances that allow one to assess associations between skewed eating patterns and specified health outcomes. Such associations would have to be followed up by other more controlled assessments.

Research Needs

Findings

There is a need, in the committee's judgment, for a broad research and technology development agenda to improve methods for predicting, identifying, and assessing unintended health effects from the genetic modification of food. An additional benefit is that the tools and techniques developed can also be applied to safety assessment and monitoring of foods produced by all methods of genetic modification.

The tools and techniques already developed can be applied to the safety assessment and monitoring of foods produced by all methods of genetic modification. However, although current analytical methods can provide a detailed assessment of food composition, limitations exist in identifying specific differences in composition and interpreting their biological significance.

Recommendation 6

A significant research effort should be made to support analytical methods technology, bioinformatics, and epidemiology and dietary survey tools to detect health changes in the population that could result from genetic modification and, specifically, genetic engineering of food. Specific recommendations to achieve this goal include:

• Focusing research efforts on improving analytical methodology in the study of food composition to improve nutrient content databases and increase understanding of the relationships among chemical components in foods and their relevance to the safety of the food.

• Conducting research to provide new information on chemical identification and metabolic profiles of new GM foods and proteomic profiles on individual compounds and complex mixtures in major food crops and use that information to develop and maintain publicly accessible databases.

• Developing or expanding profiling databases for plants, animals, and microorganisms that are organized by genotype, maturity, growth history, and other relevant environmental variables to improve identification and enhance traceability of GMOs.

• Developing improved bioinformatics tools to aid in the interpretation of food composition data derived from targeting and profiling methods.

Recommendation 7

Research also is needed to determine the relevance to human health of dietary constituents that arise from or are altered by genetic modification. This effort should include:

• Focusing research efforts on developing new tools that can be used to assess the potential unintended adverse health effects that result from genetic modification of foods. Such tools should include profiling techniques that relate metabolic components in food with altered gene expression in relevant animal models to specific adverse outcomes identified in GM animal models (animals genetically modified by contemporary biotechnology methods that are proposed to enter the food system).

• Developing improved DNA-based immunological and biochemical tags for selected GM foods entering the marketplace that could be used as surrogate markers to rapidly identify the presence and relative level of specific foods for postmarket surveillance activities.

• Developing improved techniques that enable toxicological evaluations of whole foods and complex mixtures, including:
  — microarray analysis,
  — proteomics, and
  — metabolomics.

CONCLUDING REMARKS

The committee was charged with the task of identifying appropriate scientific questions and examining methods for determining unintended changes in the levels of nutrients, toxicants, allergens, or other compounds in food from genetically engineered organisms compared with those from other genetic modification processes and outlining methods to assess the potential short- and long-term human health consequences of such changes. To address its charge, the committee took into account the current state of the science for available analytical techniques. These techniques have improved in recent decades, as has knowledge and understanding of food safety. Nevertheless, substantial gaps remain, including our ability to:

• identify compositional changes in food and other complex mixtures,
• determine the precise chemical structure of more than a small number of compounds in a tissue,
• determine the structure-function relationships between compounds in food and their relevance to human health, and
• predict and assess the potential outcome of unintended changes in food on human health.

In consideration of the advances and limitations to available analytical techniques, the committee developed an appropriate paradigm for:

• identifying appropriate comparators,
• increasing understanding of the determinants of compositional variability,

- increasing understanding of the bioactivity in humans, if any, of secondary metabolites in commonly consumed foods,
- developing more sensitive tools for assessing potential unintended health effects (e.g., in whole food), and
- improving methods for tracking and tracing exposure in GE food.

The recommendations presented in this chapter reflect the committee's application of the framework it has developed to questions of identification and assessment of unintended adverse health effects from foods produced by all forms of genetic modification, including genetic engineering and they can serve as a guide for evaluation of future technologies.

Appendixes

A

Glossary

Adventitious bacteria	Bacteria that originate from a source other than the organism or environment in which they are found and are not inherent to that organism or environment.
Agrobacterium	A naturally occurring pathogenic bacterium of plants that can incorporate a portion of a plasmid deoxyribonucleic acid (DNA) into plant cells.
Allele	One of the variant forms of a gene at a particular locus, or location, on a chromosome. Different alleles produce variation in inherited characteristics, such as blood type.
Antinutrient	A compound (in food) that inhibits the normal uptake or utilization of nutrients or that is toxic in itself.
Apoptosis	The process of cell death, which occurs naturally as a part of normal development, maintenance, and renewal of tissue in an organism.
Bacillus thuringiensis (*Bt*)	A strain of bacteria that produces a protein toxic to certain insects that cause significant crop damage. The bacteria are often used for biological pest control. The gene that codes for the toxic protein has been engineered into other soil bacteria and also directly into some crop plants.

191

Bacteriophage (phage)	A virus that infects bacteria.
Bioinformatics	The management and analysis of data (especially DNA sequence data) using advanced computing techniques.
Biolistic device	A device that bombards target cells with microscopic DNA-coated particles. Familiarly known as the Gene Gun, it was first developed in the early 1980s.
Carcinogen	A cancer producing agent or substance. A variety of chemical agents have been shown to induce malignancy in animals, but not all of them show the same capability in humans.
Carotenoids	A group of chemically similar red to yellow pigments responsible for the characteristic color of many plant organs or fruits, such as tomatoes and carrots. Carotenoids serve as light-harvesting molecules in photosynthetic assemblies and also play a role in protecting prokaryotes from the deleterious effects of light.
Cell selection	The process of selecting cells that exhibit specific traits within a group of genetically different cells. Selected cells are often subcultured onto fresh medium for continued selection.
Chitinase gene	A gene responsible for the activity of chitinase, an enzyme that breaks down chitin (a polysaccharide that gives structural strength to the exoskeleton of insects and the cell walls of fungi).
Clone	Defines both molecular, whole-animal, and plant clones; a collection of genetically identical copies of a gene, cell, or organism.
Cloning	The propagation of genetically exact duplicates of an organism by a means other than sexual reproduction, for example, the vegetative production of new plants or the propagation of DNA molecules by insertion into plasmids. Often, but inaccurately, used to refer to the propagation of animals by nuclear transfer.
Cocultivation	Growth of cultured cells together.
Comparator	A product that is compared to another product (e.g., a genetically engineered food and a non-genetically engineered food).

Competent cells	Cells (e.g., bacteria, plant, or yeast) that can take up DNA and become genetically transformed.
Control elements	DNA sequences in genes that interact with regulatory proteins (such as transcription factors) to determine the rate and timing of expression of the genes, as well as the beginning and end of the transcript.
CpG methylation	A heritable chemical modification of DNA (replacement of cytosine by 5-methyl cytosine) that, when present in a control region, usually suppresses expression of the corresponding gene.
Cross-breeding	Mating between members of different populations (lines, breeds, races, or species).
Cytoplasmic inheritance	Hereditary transmission dependent on cytoplasmic genes (genes located on DNA outside the nucleus).
Ectopic gene expression	Expression of a (trans) gene in a tissue or developmental stage when such expression is not expected.
Electroporation	Introduction of DNA into a cell mediated by a brief pulse of electricity.
Endogenous	Derived from within; from the same cell type or organism.
Epithelial cell	The tissue that forms the superficial layer of skin and some organs. It also forms the inner lining of blood vessels, ducts, body cavities, and the interior of the respiratory, digestive, urinary, and reproductive systems.
Fitness	The ability to survive to reproductive age and produce viable offspring. Fitness also describes the frequency distribution of reproductive success for a population of mature adults.
Furanocoumarins	Toxic compounds found primarily in species of the Apiaceae and Rutaceae plant families. They come in a variety of related chemical structures and have adverse effects on a wide variety of organisms, ranging from bacteria to mammals.

Galactosemia	An inherited disorder characterized by an inability of the body to utilize galactose; literally, it means "galactose in the blood."
Gamete	A mature reproductive cell capable of fusing with a cell of similar origin but of opposite sex to form a zygote from which a new organism can develop. Gametes normally have a haploid chromosome content.
Gas chromatography (GC)	An instrumental analytical technique in which the volatile components of a sample are injected into a stream of gas that flows through a long heated capillary column. The sample components separate according to their relative volatility and affinity for the interior surface of the column and are quantified as they exit the column using various types of detectors.
Gas chromatography-mass spectrometry (GC-MS)	An instrumental analytical technique in which the components of a sample enter a mass spectrometer directly following their separation by gas chromatography.
Gene expression	The conversion of the gene's nucleotide sequence into an actual process or structure in the cell. Some genes are expressed only at certain times during an organism's life and not at others.
Gene introgression	Introduction of new genes into a population by crossing between two populations, followed by repeated backcrossing to that population while retaining the new genes.
Genetically modified	Refers to an organism whose genotype has been altered and includes alteration by genetic engineering and nongenetic engineering methods.
Genetic engineering	Changes in the genetic constitution of cells resulting from the introduction or elimination of specific genes via molecular biology (i.e., recombinant DNA) techniques.
Genotype	The genetic identity of an individual. Genotype often is evident by outward characteristics.
Germline cells	Cells that contain inherited material that comes from the eggs and sperm and that might be passed on to offspring.
Glycosylated	Term used to describe a molecule that has undergone the post-translational addition of carbohydrate groups to it (e.g., glycosylated hemoglobin).

Haplotype map	An effort to find DNA landmarks that identify specific DNA sequences shared by many individuals.
Hazard	A substance or agent that, upon exposure, might result in a defined harm.
Hemosiderosis	A focal or general increase in tissue iron stores without associated tissue damage.
Herpetiformis	A chronic, extremely itchy rash associated with sensitivity of the intestine to gluten in the diet (celiac sprue).
Heterozygous	The condition in which an organism has inherited two different alleles of a specific gene pair from its parents.
Hirudin	A potent clotting inhibitor produced by leeches. The gene for this protein has been genetically engineered into canola plants.
Homocystinuria	An inherited disorder of the metabolism of the amino acid methionine; it is inherited as an autosomal recessive trait.
Homologous recombination	Rearrangement of DNA sequences on different molecules by crossing over in a region of identical sequence.
Homologs	In diploid organisms, a pair of matching chromosomes.
Homozygous	The condition in which an organism has inherited two identical alleles of a specific gene pair from its parents.
Horizontal gene transfer	Transmission of DNA involving close contact between the donor's DNA and the recipient, uptake of DNA by the recipient, and stable incorporation of the DNA into the recipient's genome.
Hybrid	Progeny of genetically different parents, usually of the same species, that has enhanced productivity over either parent. Generally, the more genetically diverse the parent lines, the more hybrid vigor, or heterosis, is observed in the hybrid progeny.
Hybridoma	A fast-growing culture of cloned cells made by fusing a cancer cell to some other cell, such as an antibody-producing cell.

Immunoglobulin E (IgE)	A component of the human immune system implicated in the expression of allergies.
Inbred line	A population of plants that are self-pollinated over several generations and are largely genetically homogenous. Such inbreeding results in a population of plants with nearly identical genetic composition, and homozygosity, or genetic uniformity, at every gene locus.
Integration	The covalent joining of a piece of DNA into genomic DNA.
Landrace	A primitive cultivar (in contrast to a named modern cultivar). Landraces of a particular crop are a collection of plants that were developed and maintained by traditional farmers. While they are genetically improved over wild versions of the species, they are not as well developed as modern commercial cultivars.
Lipofection	A method of transfection in which DNA is incorporated into lipid vesicles (liposomes), which then are fused to the membrane of the target cells.
Liquid chromatography (LC)	An instrumental analytical technique in which the soluble components of a sample are injected into a stream of liquid solvent pumped through a tube (column) packed with small retentive particles. The sample components separate according to their relative affinity for the flowing solvent (mobile phase) and the surface of the solid particles with which the column is packed (stationary phase). The sample components are quantified as they exit the column using various types of detectors. In its contemporary format, analytical liquid chromatography is usually termed high-performance liquid chromatography (HPLC).
Liquid chromatography-mass spectrometry (LC-MS)	An instrumental analytical technique in which the components of a sample enter a mass spectrometer directly following their separation by gas chromatography.
Locus	The place on a chromosome which is occupied by a gene.

Locus-control regions	Segments of DNA important for the correct and coordinated expression of large regions (such as those encoding hemoglobins).
Lysine	A basic amino acid that is produced chiefly from many proteins by hydrolysis. It is essential in the nutrition of humans and animals.
Mass spectrometry (MS)	An instrumental analytical technique in which a chemical compound is detected and identified according to the pattern of masses and abundances of charged particles obtained when that compound is subjected to ionization in an electron beam or other ionizing conditions.
Meiosis	The special cell division process by which the chromosome number of a reproductive cell becomes reduced to half (n) the diploid (2n) or somatic number.
Mesenchymal cells	Cells that form the supporting tissue of an organ or blood vessel.
Metabolomics	Systematic global analysis of nonpeptide small molecules, such as vitamins, sugars, hormones, fatty acids, and other metabolites. It is distinct from traditional analyses that target only individual metabolites or pathways.
Methylation	Addition of a methyl group ($-CH_3$) to a macromolecule, such as a specific cytosine and, occasionally, adenine residues in DNA.
Microinjection	Introduction of DNA into a cell by injection through a very fine needle.
Microprojectile bombardment	Also known as particle acceleration, or biolistic bombardment (using the "Gene Gun"), this technique is used to transform cells using small gold or tungsten particles that are coated with DNA and literally shot into a cell.
Microsatellite DNA	A small segment of DNA with a repeated sequence. This segment is made up of short nucleotide sequences, which when tagged and amplified using polymerase chain reaction (PCR; see below), can be used as markers for research purposes.
Mitochondria	Small cytoplasmic organelles that carry out aerobic respiration; oxidative phosphorylation takes place to produce adenosine triphosphate.

Mitotic division	A method of indirect division of a cell, consisting of a complex of various processes, by means of which the two daughter nuclei normally receive identical complements of the number of chromosomes characteristic of the somatic cells of the species.
Mobilization	The transfer of genes from one place to another (in the same or a different cell or organism) mediated by a retrovirus or transposable element.
Monoclonal antibody	An antibody of a single type produced by a genetically identical group of cells (clone). Usually a fusion of an antibody-producing blood cell and a cancer cell. See hybridoma.
Mutagenesis (or mutation breeding)	A process whereby the genetic information of an organism is changed in a stable, heritable manner, either in nature or induced experimentally via the use of chemicals or radiation. In agriculture, these genetic changes are used to improve agronomically useful traits.
Mycotoxins	Toxic substances of fungal origin, such as aflatoxins.
Neomycin phosphotransferase, type II	A bacterial gene encoding resistance to several common antibiotics (kanamycin, neomycin, G418), widely used as a selectable marker in eukaryotic cells.
Northern blot analysis	A technique used to analyze ribonucleic acid (RNA). RNA is separated by size, blotted (transferred onto a membrane), and then is detected by a special probe that allows information (e.g., size and abundance) about a particular species of RNA to be revealed.
Oligopeptide	A molecule consisting of a small number of amino acid units.
Oral gavage	Feedings given through a tube passed through the mouth and into the stomach.
Organoleptic	Relating to the senses (taste, color, odor, feel). For example, traditional meat and poultry inspection techniques are considered organoleptic because inspectors perform a variety of such procedures, involving visually examining, feeling, and smelling animal parts to detect signs of disease or contamination.

Paracrine	A form of signaling in which the target cell is close to the signal releasing cell. Neurotransmitters and neurohormones usually are considered to fall into this category.
Phage	See *bacteriophage*.
Pharmacogenetics	The study of how people's genetic makeup affects their response to medicines.
Phenotype/Phenotypic	The visible and/or measurable characteristics of an organism (i.e., how it appears outwardly) as opposed to its genotype, or genetic characteristics.
Phenotype Mapping	A process used to identify landmarks on DNA through functional analysis of trait genes.
Phenylketonuria	A genetic disorder in which the body cannot break down the amino acid phenylalanine.
Phytoalexins	Plant-produced substances that are toxic or inhibit the growth of microorganisms, especially phytopathogenic fungi (and can be harmful for the host plant itself). Phytoalexins are many chemically distinct chemical compounds (e.g., isoflavonids, furanocoumarins).
Plant population	In a natural, or unmanaged, environment, a plant population typically is composed of many different species, some of which can be edible and desirable to humans, and others that might not. In the context of an agricultural, or managed, environment, a plant population could be a field of plants—typically of one variety—but invariably consisting of some plants of different varieties and some different species.
Plasmid	A circular DNA molecule capable of replication in host bacteria. Plasmids are the usual means of propagation of DNA for transfection or other purposes. Plasmids are also occasionally found in certain fungi and plants.
Pleiotropy	A phenomenon whereby a particular gene affects multiple traits.
Polygenic	Refers to a trait or phenotype whose expression is the result of the interaction of numerous genes.
Polymerase chain reaction (PCR)	A method for making multiple copies of fragments of DNA. It uses a heat-stable DNA polymerase enzyme and cycles of heating and cooling to successively split apart the strands of double-stranded DNA and use the single strands as templates for building new double-stranded DNA.

Polymorphism | A natural variation in a gene, DNA sequence, or chromosome, which may not have adverse effects on the individual and occurs with fairly high frequency in the general population.

Proctocolitis (eosinophilic proctocolitis) | An inflammation of the colon and rectum, characterized by elevated levels of eosinophils (a type of white blood cell) in the tissues of these organs.

Promoter | A regulatory element that specifies the start site of transcription.

Protease | An enzyme that hydrolyzes proteins, cleaving the peptide bonds that link amino acids in protein molecules.

Proteomics | The analysis of complete complements of proteins. Proteomics includes not only the identification and quantification of proteins, but also the determination of their localization, modifications, interactions, activities, and, ultimately, their function.

Protoplast fusion | A technique in which protoplasts (plant cells from which the cell wall has been removed by mechanical or enzymatic means) are fused into a single cell.

Provirus | The integrated DNA form of a retrovirus.

Proximate analysis | The analytical determination of the major classes of food components, usually including total protein, fat, and carbohydrate, dietary fiber, water, and ash (minerals).

Psoralens | Mutagenic, carcinogenic agents that can act as phytoalexins. Psoralens are found in a variety of fruits and vegetables, such as celery and carrots, and they are known to make human skin sensitive to long-wave ultraviolet radiation.

Radioallergosorbent (test) | Allergen testing using blood samples to identify an allergen capable of causing an allergic response.

Recombinant | Refers to a genotype with a new combination of genes, in contrast to parental type.

Recombinant DNA techniques | Procedures used to join together DNA segments. Under appropriate conditions, a recombinant DNA molecule can enter a cell and replicate there.

Restriction endonuclease	An enzyme that cuts a DNA molecule at a particular base sequence.
Retroviral vectors	Vector constructs in which the internal genes of a retrovirus are replaced by the gene of interest, flanked by the viral long terminal repeats and packaging signals. After transfection of helper cells, the vector is packaged into virus particles. Infection of target cells with these particles leads to integration of the gene into cellular DNA as part of a provirus.
Retrovirus	An enveloped virus that replicates by reverse transcription of its RNA genome into DNA, followed by integration of the DNA into the cell genome to form a provirus. Expression of the provirus (as though it were a cellular gene) leads to the production of progeny virus particles.
Reverse transcription	The process of copying RNA into DNA.
Risk	The likelihood of a defined hazard being realized, which is the product of two probabilities: the probability of exposure, $P(E)$, and the probability of the hazard resulting given that exposure has occurred, $P(H/E)$ (i.e., $R = P(E) \times P(H/E)$).
Selectable marker	A gene, usually encoding resistance to an antibiotic, added to a vector construct to allow easy selection of cells that express the construct from the large majority of cells that do not.
Selection	Differential survival and reproduction phenotypes. Also, a system for either isolating or identifying specific organisms in a mixed culture; observing the characteristics of plants and choosing (selecting) to use only those organisms that have desired or superior characteristics.
Serine phosphorylation	The chemical attachment of phosphorus to a molecule of serine, an amino acid that serves as a storage source of glucose by the liver and muscles.
Servomechanism	An automatic feedback device in which the controlled variable is mechanical position or any of its time derivatives.

Sexual selection	The type of selection in which there is competition among males for mates, and characteristics enhancing the reproductive success of the carrier are perpetuated irrespective of their survival value.
Silencing	Shutdown of transcription of a gene, usually by methylation of C residues.
Solanaceous	Pertaining to plants of the family Solanaceae which includes tomato, tobacco, potato, pepper, and many weeds.
Somaclonal variation	Epigenetic or genetic changes, sometimes expressed as a new trait, resulting from in vitro culture of higher plants.
Somatic cells	Cells of body tissues other than the germline.
Southern blot analysis	A technique used to obtain information about identity, size, and abundance of a specimen of DNA. DNA fragments are transferred to membrane filters so that specific base sequences can be detected.
StarLink	A commercial brand of transgenic maize approved for animal feed only, but which also had been found in the human food supply.
Stromal cells	Nonblood cells derived from blood organs, such as bone marrow or fetal liver, which are capable of supporting growth of blood cells in vitro. Stromal cells that make this matrix within the bone marrow are also derived from mesenchymal stem cells.
Substantial equivalence	A concept that has been proposed to measure whether a biotechnology-derived food or crop shares similar health and nutritional characteristics with its conventional counterpart. The Food and Agriculture Organization of the United Nations, and the World Health Organization have attempted to develop substantial equivalence as an internationally agreed upon principle.
T-DNA	DNA encoded on a plasmid of *Agrobacterium* that integrates into the genome of a plant cell.
Telomerase	The enzyme, absent from most somatic cells but present in germline cells, that restores telomeres to their normal length.

Telomeres | The simple repeated sequences at the ends of chromosomes that protect them from loss of coding sequence during replication. In the absence of telomerase, telomeres become progressively shorter with each cell division, and this shortening is the major cause of senescence of cells in culture.

Toxicant | Any substance or material that can injure living organisms through physicochemical interactions.

Toxin | A poisonous substance (of animal, mineral, vegetable, or microbial origin) that can cause damage to any living tissues.

Transfection | Alteration of the genome of a cell by direct introduction of DNA, a small portion of which becomes covalently associated with the host cell DNA.

Transgene | A gene construct introduced into an organism by recombinant DNA methods.

Transgenic organism | An organism into which DNA has been introduced using recombinant DNA methods.

Transposase | The enzyme responsible for moving a transposon from one place to another.

Transposon | A DNA element capable of moving (transposing) from one location in a genome to another in the same cell through the action of transposase.

Vector | A type of DNA, such as a plasmid or phage, that is self-replicating and that can be used to transfer DNA segments among host cells. Also, an insect or other organism that provides a means of dispersal for a disease or parasite.

Vertical transmission | Inheritance of a gene from parent to offspring.

Virion | The extracellular form of a virus (i.e., a virus particle).

Xanthophylls | A yellow oxygen-containing carotenoid, present in chloroplasts.

SUBREPORT ON ANIMAL CLONING

Blastomere	Any one of the cells formed from the first few cleavages in animal embryology. The embryo typically divides into two, then four, then eight blastomeres, and so forth.
Casein	The primary protein in milk.
Chimeras	Animals (or embryos) composed of cells of different genetic origin.
Chromatin	The genetic material that makes up chromosomes. Specifically, it is a tangled, fibrous mixture of DNA and protein found within a eukaryotic nucleus.
Embryo rescue	A sequence of tissue culture techniques used to enable a fertilized immature embryo resulting from an interspecific cross to continue growth and development, until it can be regenerated into an adult plant.
Enucleated oocyte (cytoplast)	An egg cell from which the nucleus has been removed mechanically.
Epigenetics	The study of mechanisms that produce phenotypic effects by altering gene activity without altering the nucleotide sequence or genotype of an organism.
Hydrops fetalis	Also known as fetal hydrops, this condition occurs when there is a presence of fetal subcutaneous tissue edema (the abnormal accumulation of serous fluid), including the escape of serous fluid into one or more body cavities, resulting in a newborn suffering from severe edema.
Insulin-like growth factors	Also known as IGF-I and IGF-II, they are polypeptides that have an amino acid sequence that shares some similarity with insulin. IGF-I is synthesized in the liver and other tissues. Synthesis of IGF-I is responsive to growth hormone, and IGF-II is known for having multiplication stimulating activity.

Knock-in	Replacement of a gene by a mutant version of the same gene using homologous recombination.
Knock-out	Inactivation of a gene by homologous recombination following transfection with a suitable DNA construct.
Major Histocompatibility Complex (MHC)	A large genomic region or family of genes in most vertebrates, which contains several genes with important functions for the immune system.
Monozygotic twin	Identical twins that are the result of a single zygote (fertilized egg) splitting into two cell masses and becoming two individuals. The twins are genetically identical and are always of the same sex.
Nuclear transfer	The generation of a new animal nearly identical to another one by injection of the nucleus from a cell of the donor animal into an enucleated oocyte of the recipient.
Oocyte	The egg mother cell; it undergoes two meiotic divisions (oogenesis) to form the egg cell. The primary oocyte is before completion of the first meiotic division; the secondary oocyte is after completion of the first meiotic division.
Parturition	The process of giving birth.
Pronuclear injection	The use of a fine needle to inject DNA into the nucleus of an unfertilized egg.
Senescent cells	Animal cells that have nearly reached the limit of lifespan (usually around 50 doublings) in cell culture and are beginning to show signs of impending death.
Xenotransplantation	Transplantation of cells, tissues, or organs from one species to another.
Zygote	A fertilized oocyte

B

Open Session and Workshop Agendas

Open Session
Monday, September 23, 2002
National Academy of Sciences
Room 105
2100 C Street, NW
Washington, DC

1:00 p.m. Welcome, Introductions, and Purpose of the Public Session
Bettie Sue Masters, Committee Chair

Presentations from Representatives of the Sponsoring Agencies

1:10 U.S. Department of Agriculture
Michael Schechtman, Office of the Deputy Secretary
Charles Edwards, Food Safety and Inspection Service

2:10 Food and Drug Administration
Joseph Levitt, Center for Food Safety and Applied Nutrition
James Maryanski, Center for Food Safety and Applied Nutrition

3:10 Break

3:30 U.S. Environmental Protection Agency
J. Thomas McClintock, Office of Science Coordination and Policy

4:30   Open Discussion

4:45   Adjourn

Workshop
Tuesday, January 7, 2003
Keck Building
Room 100
500 Fifth Street, NW
Washington, DC

8:30 a.m.   Welcome and Introductory Remarks
             Bettie Sue Masters, Committee Chair

8:45   Methods for Genetically Modifying Animals and Their
             Applications
             José Cibelli, Advanced Cell Technology, Inc.

9:30   Toxicity? Transgene Expression in Muscle as a Food Product
             Robert Schwartz, Baylor College of Medicine

10:15   Break

10:30   Determining Unintended Health Effects of Biotechnology-
             Derived Foods
             Ian Munro, CanTox, Inc.

11:15   Assessing Foods Derived from Genetically Modified Crops for
             Unintended Effects—an Industry Perspective
             Roy Fuchs, Monsanto Company

12:00 noon   Lunch

1:00 p.m.   Compositional Analyses of Foods and Feeds Derived from
             Biotechnology: A Lens into Unintended Effects
             Bruce Chassy, University of Illinois, Urbana-Champaign

1:45   The Use of Profiling Methods for Identification and Assessment
             of Unintended Effects in Genetically Modified Foods
             *Harry Kuiper, RIKILT Wageningen University and Research
             Center, The Netherlands*

2:30   Break

2:45   The International Life Sciences Institute Crop Composition
            Database
            Ray Shillito, Bayer CropScience

3:30   FDA Policy for Reviewing Genetically Engineered Crops: A
            Case-Study Assessment of FDA Oversight
            *Douglas Gurian-Sherman, Center for Science in the Public
            Interest*

3:55   Public Comment Session

4:20   Closing Remarks
            Bettie Sue Masters, Committee Chair

4:30   Adjourn

C

Committee Member Biographical Sketches

Bettie Sue S. Masters (*chair*) is the Robert A. Welch Foundation Professor in Chemistry in the Department of Biochemistry at the University of Texas Health Science Center at San Antonio. Her research involves the determination of the structure and function relationships of heme- and flavin-requiring enzymes, specifically nitric oxide synthases and NADPH-cytochrome P450 reductase, employing a variety of spectroscopic and crystallographic techniques. Currently, she serves on the Advisory Committee to the Director of the National Institutes of Health and she is president of the American Society for Biochemistry and Molecular Biology. Dr. Masters received her Ph.D. in biochemistry from Duke University. She is a member of the Institute of Medicine.

Fuller W. Bazer is associate vice chancellor and associate director of the Texas Agricultural Experiment Station and executive associate dean in the College of Agriculture and Life Sciences at Texas A&M University. He is also professor and holds the O.D. Butler Chair in the Department of Animal Sciences, with joint appointments in the Departments of Veterinary Anatomy and Public Health, Veterinary Physiology and Pharmacology, and Human Anatomy and Medical Neurobiology. His research, which is partially funded by the U.S. Department of Agriculture, focuses on reproductive and developmental biology, specifically the molecular and cellular mechanisms of pregnancy recognition signals from the conceptus to the uterus, and on uterine biology in domestic animals. Dr. Bazer is a member of the Scientific Advisory Board of ADViSYS, Inc., which is developing novel approaches to modulate growth hormone in animals for therapeutic and performance enhancement applications. He earned his Ph.D. in animal science at North Carolina State University.

Shirley A. A. Beresford is a professor in the Department of Epidemiology of the School of Public Health and Community Medicine at the University of Washington. She is also a member in the Epidemiology and Cancer Prevention Research Programs at Fred Hutchinson Cancer Research Center. Dr. Beresford's research interests are in the areas of nutritional epidemiology and cancer prevention. She is chair of the Dietary Modification Advisory Committee for the Women's Health Initiative. She received her Ph.D. in epidemiology from the University of London.

Dean DellaPenna is a professor in the Department of Biochemistry and Molecular Biology at Michigan State University. He uses molecular, genetic, and biochemical approaches to understand fundamental processes, including secondary metabolic pathways, in plants of importance to agriculture. His research is partially funded by the U.S. Agency for International Development and the National Science Foundation. Dr. DellaPenna holds several patents on biochemical and genetic techniques related to his research. He is a consultant and a member of a scientific advisory board for a genomics-based pharmaceutical company. Dr. DellaPenna received his Ph.D. in plant physiology from the University of California at Davis.

Terry D. Etherton is a distinguished professor of animal nutrition and head of the Department of Dairy and Animal Science at The Pennsylvania State University. His research focuses on endocrine regulation of animal growth. He is an authority on the role of agricultural biotechnology in food production systems. Dr. Etherton chaired the National Research Council committee that authored the report, Metabolic Modifiers: Effects on the Nutrient Requirements of Food-Producing Animals. He is co-chair of the Federation of Animal Science Societies Scientific Advisory Committee on Biotechnology. Dr. Etherton received his Ph.D. in animal sciences from the University of Minnesota.

Cutberto Garza is a professor in the Division of Nutritional Sciences at Cornell University. His research interests are maternal and infant nutrition, nutrient needs of women and young children, and long-term metabolic consequences to perinatal nutrition. Dr. Garza chairs an international effort sponsored by the World Health Organization (WHO), the United Nations University, and others to develop new international references for child growth and chairs the Technical Advisory Group for the Review of Safety, Nutritional Value, Suitability, and Appropriateness of Foods used by the World Food Program. Previously, he cochaired a review of genetically modified foods for the United States and the European Union and chaired the 2000 U.S. Dietary Guidelines Committee for the U.S. Department of Agriculture and the Department of Health and Human Services. He is a member of the the WHO Reference Group on a Global Strategy on Diet, Physical Activity, and Health, and until recently, served on the National Institutes of Health Fogarty Center's Advisory Board. He is a member of the Institute of

Medicine and former chair of the Food and Nutrition Board. Dr. Garza received his M.D. from Baylor College of Medicine and his Ph.D. from the Massachusetts Institute of Technology.

Lynn R. Goldman is a professor in Environmental Health Sciences and Health Policy and Management at the Johns Hopkins University Bloomberg School of Public Health. She is also director of the Mid-Atlantic Public Health Training Center and a visiting scientist at the Centers for Disease Control and Prevention. Dr. Goldman's expertise is in environmental epidemiology and toxicology. She previously as served as division chief of Environmental and Occupational Disease Control in the California Department of Health Services and then as assistant administrator in the Office of Prevention, Pesticides and Toxic substances, at the Environmental Protection Agency, where she led the pollution prevention, pesticide regulation, toxic substances control, and right-to-know programs.

Sidney Green, Jr. is a graduate professor in the Department of Pharmacology at Howard University. Previously, he held several positions within the Food and Drug Administration's Division of Toxicological Research, including chief of the Whole Animal Toxicology Branch, associate director for Laboratory Investigations, and director of the Division. Dr. Green also was a branch chief for the Environmental Protection Agency and director of toxicology for a private testing laboratory. Dr. Green has broad knowledge of the field of toxicology; specific areas of expertise include food toxicology, genetic toxicology, and the use of alternatives to whole animals in toxicology. He is a fellow of the Academy of Toxicological Sciences. Dr. Green received his Ph.D. in biochemical pharmacology from Howard University.

Jesse F. Gregory, III is a professor in the Department of Food Science and Human Nutrition at the University of Florida. His research interests are in the areas of food chemistry and nutritional biochemistry, specifically on the bioavailability, metabolism, and function of B vitamins, with primary emphasis on folate and vitamin B_6. Dr. Gregory received his Ph.D. in food science from Michigan State University.

Jennifer Hillard is a volunteer with the Consumer Interest Alliance. From 1996 to 2002, she served as National Vice President of Issues and Policy at the Consumer Association of Canada (CAC). Her responsibilities involved coordinating CAC's work on a range of issues, including those related to food, health, and biotechnology. She has produced informational booklets in collaboration with the CAC and the Food Biotechnology Communications network, a voluntary organization of federal and provincial governmental departments and associations from the biotechnology and food producer industries. She has also written many health and safety articles for publications designed for low literacy consumers. Ms. Hillard also served as acting chair of CAC's Food and Agriculture Committee and is a member of a multi-stakeholder advisory committee to the Canadian Research Institute for Food Safety at the University of Guelph, the Canadian General Standards Board Committee (this multi-stakeholder committee is developing a standard for voluntary labeling of food produced from biotechnology), and the Canadian Biotechnology Advisory Committee-Genetically Modified Food Reference Group. She has participated in consultations on various aspects of agricultural biotechnology with Health Canada, Industry Canada, Environment Canada, and Agriculture and Agri Food Canada.

Alan G. McHughen is a biotechnology specialist in the Department of Botany and Plant Sciences at the University of California, Riverside. Until recently, he was professor and senior research scientist in the Crop Development Centre in the College of Agriculture at the University of Saskatchewan. Dr. McHughen's research interests include crop improvement using both conventional breeding and genetic engineering techniques. He helped develop Canada's regulations covering the environmental release of plants with novel traits. His recent book, *Pandora's Picnic Basket: The Potential and Hazards of Genetically Modified Foods*, describes in layperson terms the technologies underlying genetic modification of foods. Dr. McHughen received his D. Phil. from Oxford University.

Sanford A. Miller is a senior fellow and adjunct professor at the Center for Food and Nutrition Policy at Virginia Polytechnic and State University in Alexandria, Virginia. Previously, he was director of the Center for Food Safety and Applied Nutrition at the Food and Drug Administration (1978–1987), and dean of the Graduate School of Biomedical Sciences and a professor in the Departments of Biochemistry and Medicine at the University of Texas Health Science Center at San Antonio (1987–2000). Dr. Miller's research interests are in the areas of food safety and public policy, nutrition, and food toxicology. He is the chair of the U.S. Food and Drug Administration's Food Advisory Committee. He previously served on the Institute of Medicine's Food and Nutrition Board and is a member of the National Research Council's standing Committee on Agricultural Biotechnology, Health, and the Environment. Dr. Miller received his Ph.D. from Rutgers University.

Stephen L. Taylor is the Maxcy Distinguished Professor and head of the Department of Food Science and Technology and director of the Food Processing Center at the University of Nebraska. His research involves food allergies and allergy-like diseases, the assessment of the allergenicity of genetically-modified foods, and the development of immunochemical methods for the detection of allergens, proteins, and toxins. He has served as a consultant to a number of food and biotechnology companies and also serves on the Advisory Board for the Joint Institute of Food Safety and Applied Nutrition, sponsored by the University of

Maryland and the Food and Drug Administration, and is codirector of the Food Allergy Research and Resource Program at the University of Nebraska-Lincoln. He is a member of the Institute of Medicine's Food and Nutrition Board and served as chair of its Food Chemicals Codex committee from 1989–2000. Dr. Taylor is an internationally recognized expert on food allergens and food safety, particularly as they relate to bioengineered foods. He received his Ph.D. in biochemistry from the University of California, Davis.

Timothy R. Zacharewski is an associate professor in the Department of Biochemistry and Molecular Biology and in the National Food Safety and Toxicology Center at Michigan State University. His research interest is in the area of mechanistic toxicology; specifically, he studies how synthetic and natural chemicals elicit toxicity by altering gene expression. His laboratory utilizes genomic (including microarray), biochemical, and toxicological techniques. Dr. Zacharewski received his Ph.D. in toxicology from Texas A&M University.

Subreport

Methods and Mechanisms of Genetic Manipulation and Cloning of Animals

As part of its charge, the committee was asked to prepare a subreport evaluating methods for detecting potential unintended compositional changes across the spectrum of messenger ribonucleic acid (mRNA), proteins, metabolites and nutrients that may occur in food derived from cloned animals that have not been genetically modified via genetic engineering methods. Detailed descriptions of methods used in animal cloning and biotechnology are provided in the report *Animal Biotechnology: Science-Based Concerns* (NRC, 2002). In addition, the committee was charged with evaluating methods to detect potential, unintended, adverse health effects of foods derived from cloned animals.

INTRODUCTION

Since the onset of modern biotechnology, scientists have made discoveries leading to the development of new techniques for animal agriculture. Applications of biotechnology to animal agriculture include improving milk production and composition; increasing growth rate of meat animals; improving productive efficiency, or gain-to-feed ratios, and carcass composition; increasing disease resistance; enhancing reproductive performance; increasing prolificacy; and altering cell and tissue characteristics for biomedical research and manufacturing. Continued development of new biotechnologies also will allow farm animals to serve as sources of both biopharmaceuticals for human medicine and organs for transplantation.

ANIMAL BIOTECHNOLOGY

A number of animal biotechnologies have already been developed and are in commercial use. One such example is recombinant-derived bovine somatotropin (bST). Recombinant bST has been approved by the Food and Drug Administration for use in the U.S. dairy industry and is also approved for use by 18 other countries (CAST, 2003; Etherton and Bauman, 1998). Commercial use in the United States began in early 1994 and has increased to the point that about one-half of all U.S. dairy herds, comprising more than 3 million cows, are receiving bST (Bauman, 1999). Milk yield increases in response to bST typically range from 10 to 15 percent (about 4–6 kg/d), although larger increases may occur when the management and care of the animals are excellent (Bauman, 1992; Chillard, 1989; NRC, 1994). It is, however, important to distinguish the use of bST from other biotechnologies, such as transgenic or cloned animals. Application of recombinant bST is a biotechnology in which a recombinant-derived protein is administered by injection to the recipient animal without changing the animal's genetic composition or genome.

The application of genomics—the study of how the genes in deoxyribonucleic acid (DNA) are organized and expressed—and bioinformatics in animal agriculture will provide new genetic markers for improved selection for desired traits in all livestock species. Transgenic biology provides a means of altering animal genomes to achieve desired production and health outcomes of commercial value and societal importance. For example, genetic modification of animals may lead to technologies that reduce the major losses that occur during the first months of embryogenesis. Biotechnology also offers potential to animal agriculture as a means to reduce nutrients and odors from manure and volume of manure produced, resulting in animals that are more environmentally friendly (CAST, 2003).

The advent of techniques to propagate animals by nuclear transfer, also known as cloning, potentially offers many important applications to animal agriculture, including reproducing highly desired elite sires and dams. Animals selected for cloning will be of great value because of their increased genetic merit for increased food production, disease resistance, and reproductive efficiency, or will be valued because they have been genetically modified to produce organs for transplantation or products with biomedical application.

Before entering the marketplace, new agricultural biotechnologies are evaluated rigorously by the appropriate federal regulatory agencies to ensure efficacy, consumer safety, and animal health and well-being. The development of technologies to clone animals used for food production has raised the question of whether there are unintended compositional changes in food derived from these animals that may, in turn, result in unintended health effects.

CLONING

Cloning, a term originally used primarily in horticulture to describe asexually produced progeny, means to make a copy of an individual organism, or in cellular and molecular biology, groups of identical cells and replicas of DNA and other molecules. For example, monozygotic twins are clones. Although *clone* is descriptive for multiple approaches for cloning animals, in this report *clone* is used as a descriptor for somatic cell nuclear transfer.

Animal cloning during the late 1980s resulted from the transfer of nuclei from blastomeres of early cleavage-stage embryos into enucleated oocytes, and cloning of livestock and laboratory animals has resulted from transferring a nucleus from a somatic cell into an oocyte from which the nucleus has been removed (Westhusin et al., 2001; Wilmut et al., 1997).

Somatic cell nuclear transfer can also be used to produce undifferentiated embryonic stem cells, which are matched to the recipient for research and therapy that is independent of the reproductive cloning of animals. The progeny from cloning using nuclei from either blastomeres or somatic cells are not exact replicas of an individual animal due to cytoplasmic inheritance of mitochondrial DNA from the donor egg, other cytoplasmic factors that may influence "reprogramming" of the genome of the transferred nucleus, and subsequent development of the cloned organism (Cummins, 2001; Jaenisch and Wilmut, 2001).

Cloning by nuclear transfer from embryonic blastomeres (Willadsen, 1989; Willadsen and Polge, 1981) or from a differentiated cell of an adult (Kuhholzer and Prather, 2000; Polejaeva et al., 2000; Wilmut et al., 1997) requires that the introduced nucleus be reprogrammed by the cytoplasm of the egg and direct development of a new embryo, which is then transferred to a recipient mother for development to term. The offspring will be identical to their siblings and to the original donor animal in terms of their nuclear DNA, but will differ in their mitochondrial genes; variances in the manner nuclear genes are expressed are also possible.

Epigenetic Change in the Genome

Epigenetics is the study of factors that influence behavior of a cell without directly affecting its DNA or other genetic components. The epigenetic view of differentiation is that cells undergo differentiation events that depend on correct temporal and spatial repression, derepression, or activation of genes affecting the fate of cells, tissues, organs, and ultimately, organisms. Thus epigenetic changes in an organism are normal and result in alterations in gene expression. For example, epigenetic transformation of a normal cell to a tumor cell can occur without mutation of any gene.

Effects of Cloning

Santos and colleagues (2003) have reviewed current thinking on epigenetic marking that correlates with developmental potential of cloned bovine preimplantation embryos. They indicate that reprogramming of DNA methylation affects the entire genome of mammalian embryos both during phase I germline development when DNA methylation imprints are eliminated and during phase II preimplantation development of mammalian embryos.

Phase II DNA reprogramming is initiated upon fertilization of the oocyte for remodeling of chromatin in the male pronucleus and selective demethylation of its DNA, while subsequent DNA demethylation in early cleavage stage embryos is passive. At the blastocyst stage, de novo methylation is lineage-specific as the inner cell mass, or embryonic disc, becomes highly methylated and trophectoderm becomes hypomethylated. These epigenetic reprogramming events appear to be deficient in cloned embryos that have abnormal patterns of DNA methylation and gene expression.

Cloning by nuclear transfer using present methods is very inefficient. This is likely due to the limited ability of oocyte cytoplasm to reprogram somatic cell nuclear DNA, whereas it readily reprograms sperm cell nuclear DNA (Wilmut et al., 2002). Consequently, there are very high rates of embryonic, fetal, perinatal, and neonatal deaths, as well as birth of offspring with various abnormalities. These losses are assumed to result from inappropriate expression of genes during various stages of development, with less than 4 percent of embryos reconstructed using adult or fetal somatic cell nuclei being born as live young.

A very high percentage of fertilized eggs that develop and survive beyond the first 30 to 60 days of gestation in sheep and cattle have abnormal placental morphology, such as reduced vascularity and abnormal or few cotyledons; abnormal placental functions, such as hydroallantois and hydroamnios; and abnormal fetal development, such as enlarged liver, hydrops fetalis, dermal hemorrhaging, and swollen brain. Neonates also often experience respiratory distress and cardiovascular abnormalities (De Sousa et al., 2001; Hill et al., 1999, 2000). In addition, there is evidence for abnormalities of the immune system, brain, and digestive system of cloned animals. Mice cloned from cumulus cells become obese, but this trait is not heritable after sexual reproduction, which is direct evidence that obesity in the clones result from epigenetic events (Tamashiro et al., 2002; Wilmut et al., 2002).

Wilmut and colleagues (2002) have reported that a number of cellular factors influence the outcome of cloned offspring. First, normal development depends on the embryo having normal ploidy, which is achieved by coordinating stage of cell cycle of the recipient oocyte and donor nucleus. For donor cells, the $G_2/M/G_1/G_0$ phases of the cell cycle are associated with effective nuclear reprogramming; however, there are also conflicting results as to whether there is an advantage for donor cell nuclei to be in G_1 versus G_0.

Donor cells at metaphase seem to be most compatible with metaphase II oocytes with respect to development of embryos to the blastocyst stage, but high rates of mortality occur later. This mortality likely reflects inappropriate expression of genes at different stages of development, which is lethal. For example, perturbation of expression of H10 and/or insulin-like growth factor II (IGF-II), as well as reduced expression of IGF-II receptor in liver, kidney, heart, and muscle is associated with large lamb syndrome following embryo culture.

Changes in expression of the IGF-II receptor gene have been associated with loss of methylation at a differentially methylated region involved with genetic imprinting. There are also reports of improper expression of both imprinted and nonimprinted genes, inappropriate reactivation of the inactive X chromosome (Xue et al., 2002), and reprogramming of telomerase activity (Rideout et al., 2001) to restore telomere length in cloned animals.

Inappropriate gene expression can result from defects in the organization of nuclear material, chromatin structure, and/or activity of regulatory molecules. Chromatin proteins regulate access of transcription factors to chromosomal DNA and expression patterns for individual genes. Linker histone H1 and somatic cell histone H1 are important regulators of gene expression. Their expression changes from absent to very low during early embryonic development to low to moderate at the time of activation of the embryonic genome, and somatic cell histone H1 is lost from most mouse nuclei soon after transfer depending on cell cycle stage for donor and recipient cells. Enzymatic modifications of histones include phosphorylation, methylation, acetylation, and ubiquitination or removal of these modifications, which lead to epigenetic regulation of gene expression.

Expression of genes in higher organisms is regulated by histones H2A, H2B, H3, and H4, as well as histone H1 and other nonhistone proteins, which result in the condensed state DNA that is inaccessible to transcription factors. Transcription of genes requires "unraveling" of DNA and a higher order chromatin structure. This is achieved by a family of adenosine triphosphate-hydrolyzing enzymes that remodel chromatin by shuffling nucleosomes and a family of enzymes that modify histones covalently by acetylation, phosphorylation, methylation, or ubiquination of histone tails. If this does not occur, the genes are silent. For details on histone methylation and gene transcription, refer to Kouzarides (2002).

Widespread disruptions in DNA methylation occur in cloned embryos of mice and cattle, with abnormally high rates of methylation being retained through several cell divisions. In pigs, changes in methylation are initially similar to that for embryos resulting from in vitro fertilization, but by the four- to eight-cell stage, embryonic DNA is more extensively methylated than for embryos resulting from oocytes fertilized in vivo.

Santos and colleagues (2003) note that epigenetic reprogramming is severely deficient in cloned bovine embryos and involves histone H3 lysine 9 (H3-K9). In control bovine embryos, DNA methylation is reduced between the two- and four-cell stages followed by de novo methylation after the eight-cell stage. There is close correlation between H3-K9 methylation and overall DNA methylation. DNA methylation is up to tenfold greater at all stages of development in cloned bovine embryos, indicating that more genes are silent. Further, H3-K9 can be acetylated, indicating active gene expression, and cloned embryos are also hyperacetylated. These results suggest failure to both activate and to silence genes appropriately in cloned bovine embryos.

In normal blastocyts, there is lineage-specific hypermethylation of the inner cell mass compared with trophectoderm, but this asymmetry is not found in cloned embryos. Differences in acetylation of H3-K9 were also found, but differences were not as dramatic as those for methylation. Failure of trophectoderm to be hypomethylated may explain the aberrant placental development that is characteristic of cloned animals.

Nuclei from bovine granulosa cells yield a higher percentage of cloned blastocysts than nuclei from fetal fibroblasts and, as predicted, granulosa cell-derived clones had a higher proportion of normal DNA and histone methylation. One example of inappropriate expression of a gene is major histocompatibility (MHC) class I antigens that are not expressed by normal bovine trophectoderm/chorion. However, major histocompatibility class I genes are expressed by trophectoderm/chorion of cloned conceptuses (Hill et al., 2002), which may put the cloned conceptus at risk for rejection by the maternal immune system.

Epigenetic programming is a normal part of development, including differentiation of primordial germ cells, sperm, oocytes, and conceptus tissues. This programming allows for both temporal and spatial repression and derepression of genes and development of healthy offspring. There is no evidence that the basic genome of cloned animals differs from that of the donor of cells used for somatic cell nuclear cloning.

While most cloned animals do not survive, there is considerable evidence that a very low percentage of cloned embryos do survive because of inappropriate silencing or over-expression of genes due to abnormalities such as genome methylation, histone assembly into nucleosomes, or chromatin remodeling by linker histones, polycomb group proteins, nuclear scaffold proteins, and transcription factors. Those clones that do survive beyond the neonatal period have an apparently normal phenotype. In either case, the basic genome of the cloned embryo is not modified. However, temporal and spatial aspects of gene expression during the course of development are apparently flawed in cloned embryos and the severity of these epigenetic events likely dictate the fate of clone.

Little evidence is available in the scientific literature to assess whether the progeny of cloned animals are at increased risk for inherited or developmental defects. The recently released 2003 Food and Drug Administration (FDA) draft report *Animal Cloning: A Risk Assessment* makes the following statement about the progeny of clones: "The underlying biological assumption for progeny animals is that generation of the cells that ultimately become ova and sperm naturally resets the epigenetic signals for gene expression. This process is thought to effectively "clear" the genome of incomplete or inappropriate signals. The data to confirm this under-lying assumption are limited but consistent across species" (FDA, 2003).

EVALUATING METHODS TO DETECT POTENTIAL UNINTENDED COMPOSITIONAL CHANGES AND ADVERSE HEALTH EFFECTS OF FOODS DERIVED FROM CLONED ANIMALS

Background

Historically, equivalence of tissue or food composition has been an impor-tant component of the regulatory process to evaluate food safety (CAST, 2001; Falk et al., 2002; Juskevich and Guyer, 1990). For genetically modified (GM) plants and the animal biotechnologies reviewed by FDA, the evaluation has in-cluded comprehensive compositional analyses of plants, tissues, and milk when appropriate. The committee found that a comparable approach for products from cloned animals—primarily meat and milk—would be an appropriate, scientifi-cally based method for assessing compositional equivalence. Implicit to such as-sessments is that no increased health risk would be expected if the compositional analyses of animal products from cloned and noncloned animals were substan-tially equivalent.

Substantial Equivalence

Establishing equivalence of composition is evidence that substantive compo-sitional changes did not occur in the animal as the result of the genetic modifica-tion event. Based on studies with GM plants, substantial equivalence is analogous to "as safe as its conventional counterpart" (CAST, 2001). The approach of sub-stantial equivalence, however, is not absolute. Numerous biotechnology ap-proaches developed in animal agriculture have been to intentionally design ge-netic modification to effect changes in the composition of target nutrients or other molecules (for review, see Karatzas, 2003; Niemann and Kues, 2000). Thus the process of assessing compositional equivalence needs to be undertaken and ac-commodated for with this in mind.

For example, cloned transgenic cows have been developed that produce milk with a marked increase in ß-casein and *k*-casein (Brophy et al., 2003). Transgenic pigs have been produced that overexpress the bGH gene, which is associated with a dramatic reduction in carcass fat (85 percent reduction) and constituent fatty acid classes (Solomon et al., 1994). Genetic modification to change levels of selected nutrients in plants and animals has been, and is, an important objective of genetic engineering strategies to create designer foods (CAST, 2003; Falk et al., 2002). From the perspective of modifying the nutrient profile of foods, this has been done to increase beneficial nutrients or to decrease nutrients associated with adverse health effects, such as saturated fatty acids.

Safety Assessment

There is a long history of assessing the safety of foods introduced into the marketplace, involving an integrated multidisciplinary approach that incorporates molecular biology, protein chemistry and biochemistry, food chemistry, nutritional sciences, and toxicology. It is important to appreciate that absolute safety is not the objective with respect to any methodology or combination of methodologies used to evaluate complex substances such as food. The standard that customarily has been applied is that the food under evaluation *should be as safe as* an appropriate counterpart that has a long history of safe use. This comparative evaluation process is the foundation of establishing substantial equivalence of the food being evaluated. It also is important to emphasize that the food product itself, rather than the biotechnology process used to generate GM animals and cloned animals, should be the focal point of the evaluation. The primary objective of the safety review is to assess food safety; embedded in this is whether the process might affect the food. In addition, it is important to recognize that a statistically significant difference in one or more compounds in the food evaluated and the appropriate comparator does not necessarily imply an outcome with respect to human health. This must be evaluated on a case-by-case basis as part of the regulatory framework.

The current regulatory view of FDA is that gene-based modification of animals for food production falls under the Center for Veterinary Medicine regulations as new "animal drugs." Since epigenetic changes in the genome may lead to changes in expression of one or more genes in a manner that may be analogous to gene expression changes observed in transgenic animals, the committee determined that cloned animals should initially be evaluated in a manner that is comparable to that for animals in which genetic engineering has been used to make specific genetic modifications.

The committee also determined that cloned animals developed from transgenic parent stock for the purpose of producing pharmaceutical compounds, biomaterials, and other products not related to food production, not be allowed to enter the food chain. The committee believes this approach reduces the regulatory burden of evaluating novel biotechnology products designed for purposes other than food production.

The committee further determined that epigenetic events are normal with respect to germ cells and conceptus development, but as the result of cloning there may be changes in the extent of epigenetic events that influence expression of one or more genes. Consequently, an integral component of our evaluation was to assess the existing and evolving methods available to assess changes, both targeted and globally, in gene expression.

If a change in gene expression or alteration in mRNA abundance does occur as the result of cloning, this would be expected to result in "downstream" changes in the level of a protein encoded for by the particular mRNA species. Depending

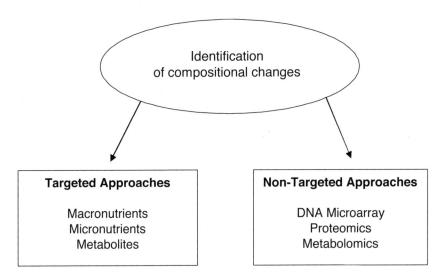

FIGURE 1 Assessment of compositional changes of food.
Adapted from Kuiper et al. (2003).

upon the biological role of the protein, this could result in a change in the abundance of a cell constituent, such as a metabolite, protein, lipid, or carbohydrate. Thus quantifying whether any of these constituents are changed in a food is important to determine whether epigenetic modification is associated with any unintended compositional change. Consequently, methodologies that are currently used or are under development to detect changes in mRNA, protein, metabolites, and nutrients will be evaluated.

Evaluation of Unintended Composition Changes

Various analytical methods can be utilized to identify compositional changes in food, irrespective of whether these are intended or unintended (Kuiper et al., 2003; also discussed in detail in Chapter 4 of the main report). These strategies can be broadly categorized as either targeted or nontargeted (see Figure 1).

Targeted Approach

Historically, the targeted analytical approach has been an important component of the FDA review process. In this approach, known individual compounds, such as nutrients, are quantified (see Box 1a and 1b for representative listing of important macronutrients and micronutrients).

Box 1a   List of possible macronutrients to quantify

Fatty acids, saturated
C4:0
C6:0
C8:0
C10:0
C12:0
C14:0
C15:0
C16:0
C17:0
C18:0
C20:0
C22:0
C24:0

Fatty acids, monounsaturated
C14:1
C16:1
C18:1
C20:1
C22:1

Fatty acids, polyunsaturated
C18:2
C18:3
C18:4

C20:4
C20:5
C22:5

Amino acids
Alanine
Arginine
Asparagine
Aspartic Acid
Cysteine
Glutamine
Glutamic Acid
Glycine
Histidine
Isoleucine
Leucine
Lysine
Methionine
Phenylalanine
Proline
Serine
Threonine
Tryptophan
Tyrosine
Valine

While analysis of specific, known compounds is important, the targeted approach may miss some unexpected changes because the list of compounds selected for assay is not all-inclusive. For GM crops, the Organization for Economic Cooperation and Development has published a list of analytes to quantify in assessing compositional changes (OECD, 2001). For GM plants, the analyses have provided information on macronutrients, micronutrients, antinutritive factors, and naturally occurring toxins. For cloned animals, it will be important to establish a comparable list. Box 1a and 1b present proposed lists of nutrients that could be quantified as part of the compositional equivalence determination for cloned animals.

Box 1b   List of possible micronutrients to quantify

Minerals	Vitamins
Calcium, Ca	Vitamin C, ascorbic acid
Iron, Fe	Thiamin
Magnesium, Mg	Riboflavin
Phosphorus, P	Niacin
Potassium, K	Pantothenic acid
Sodium, Na	Vitamin B-6
Zinc, Zn	Folate
Copper, Cu	Vitamin B-12
Manganese, Mn	Vitamin A
Selenium, Se	Vitamin E
	Vitamin D

Nontargeted Approach

Because of the vast number of individual constituents found in animal products, quantifying all of the macronutrients and micronutrients found in various tissues and in milk as a means to determine compositional equivalence is neither feasible nor practical. To enhance the likelihood of detecting unintended changes, nontargeted analytical approaches, or profiling methods such as DNA array, proteomics, and metabolomics, have been proposed as an approach for characterizing changes in GM plants (Kuiper et al., 2003; Schilter and Constable, 2002; also see Chapter 5 of the main report). The use of these profiling methodologies has the potential to provide a more extensive or global quantification of mRNA, protein, and metabolites to determine whether changes in one or more have occurred as the result of the cloning procedure.

Integration of Targeted and Nontargeted Approaches

Integration of both targeted and nontargeted approaches is a promising means to assess whether cloning has induced unintended changes in composition of meat or milk. Figure 2 presents a flow chart illustrating sequential and parallel assay steps that could be part of the profiling approach for detecting unintended effects. It must be emphasized, however, that both this report and information in the committee's main report underscore the need to establish and validate the requisite methodological specificity and sensitivity of the various nontargeted profiling methods before they can be used in practice to reliably determine whether

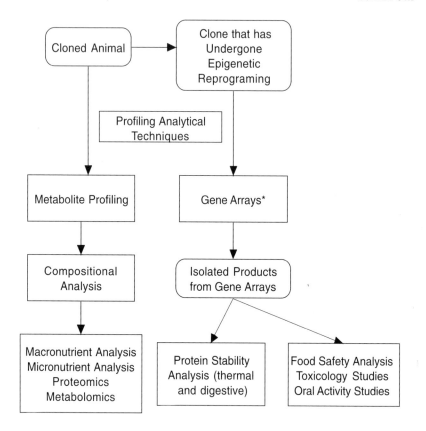

* Gene arrays identify changes in gene expression; however, there are many genes for which there is limited understanding of their biology.

FIGURE 2 Flow chart for conducting profiling analytical techniques.

unintended changes have occurred as the result of cloning (for review, see Kuiper et al., 2003).

Regardless of the analytical approaches taken to establish compositional equivalence, the following two issues must be addressed.

• Defining threshold criteria for concluding that a significant change has occurred in a constituent of an animal product *and* whether this change is associated with an unintended (and adverse) health effect in humans; and
• Standardization of sampling procedures for profiling techniques and

interlaboratory testing and assay validation (including assay sensitivity) among laboratories.

Comparator Databases

In addition to validating profiling methodologies as reliable analytical approaches to evaluate compositional equivalence, it is important to establish, validate, and utilize databases containing detailed information about normal variations in mRNA, protein, and metabolite profiles within a species. These comparator databases should account for the normal variation within an animal species that occurs in composition during growth and development, different physiological states—such as lactating versus nonlactating—and different animal management and rearing practices.

Consequently, the committee found that appropriate comparator databases could be derived from compositional data for the tissue or product produced by the species used for cloning during the time a particular species is used for productive purposes. In this example, the comparator database is not derived from animals that have been genetically modified by the application of transgenic biology.

Assessment of Methods to Detect Unintended Health Effects

Because each cloned animal is expected to be unique from an epigenetic basis, the profile of nutrients and metabolites to be assayed may need to be established on a case-by-case basis. The committee found that cloning creates a potential for temporal and spatial changes in gene expression due to epigenetic events, but without changing the basic genome of any species. There is no scientific evidence that cloning is associated with any unintended compositional change that results in an unintended health consequence in humans.

Since FDA treats genetic modification of animals as a new animal drug, it is beneficial to use the review process for recombinant bST as a case study to comment on methods of approach for evaluating health effects of foods derived from cloned animals. Bovine somatotropin was the first product of biotechnology approved for commercial use in animal agriculture. The approach for supplementing cows with bST is to provide the recombinant protein by injection.

Administration of bST increases blood levels of the protein leading to the consequent biological effects. The biological cascade invoked by administration of a protein, such as bST, is comparable to the situation where overexpression of a gene occurs, resulting in an increase in the level of a gene product due to epigenetic events. In turn, the increased level of the gene product increases the target protein in the blood and tissues with the attendant biological effect.

An integral component of the FDA review process is to establish food safety of the compound undergoing investigation. In the case of cloned animals, there is no compound per se to review since no single drug or compound is being administered to the target animal. Rather, any change that might occur as the result of cloning would be alterations in expression of one or more genes in its basic genome.

The uncertainty of what gene or genes might be affected at the expression level by cloning greatly hinders the use of any targeted approach for selecting a specific gene or few genes a priori as a means to identify an unintended effect in gene expression. Consequently, nontargeted profiling methods that provide a more global assessment of compositional changes are needed. As previously discussed, however, the nontargeted profiling techniques are not yet appropriate to use as a reliable and reproducible means to assess unintended changes in mRNA, protein, or metabolite levels. The committee concluded it is important that research be supported to evolve and refine the nontargeted profiling methodologies that, as mentioned previously, include DNA microarray, proteomics, and metabolomics.

In the future, using profiling methods to establish which genes are expressed differently will be important to not only to establish identity of the gene, but also to help establish the knowledge base about the biology of the gene product. As discussed above for gene products for which there is an understanding of the biology, safety evaluations could be made of a known gene product in a manner consistent with present FDA guidelines. However, for many gene products there will be little or no information about the biology. This is problematic because of the time and cost associated with establishing a sufficient scientific understanding of the biology.

In the case of a known gene product, FDA's review of bST provides a case study of what methods can be used to evaluate food safety. For a protein like bST, an understanding of the chemical nature, biological activity, and potential for harmful residues is important (Juskevich and Guyer, 1990). A determination of oral activity of the protein "drug" is initially required. The design of the oral toxicity studies is based on the known biological activities of the protein.

For bST, FDA required that rats be treated orally with "up to 100 times or more the dose of bST administered daily to cattle (on the basis of mg/kg of body weight) for at least 14 days" (Juskevich and Guyer, 1990). In this case, a known gene product, recombinant bST, was available for oral feeding studies.

The approach of feeding a food product derived from cloned animals to meet the 100-fold food safety margin is not feasible. Feeding this large quantity of a single food, which at some point on the dose-response curve the test food would be the only dietary source, to the test animal, may result in associated problems caused by nutrient deficiencies that may occur as the result of one food being predominant in the diet. This could be interpreted as an unintended effect when, in fact, the effect observed reflects inadequate intake of an essential nutrient.

It is important to identify each gene product affected by cloning to establish biological function of the gene product. The gene product, as a defined chemical, could be evaluated for hazard potential using conventional toxicology approaches in a way approved for bST (see Juskevich and Guyer, 1990). If expression of numerous genes is affected by the cloning event, this evaluation will be challenging, costly, and burdensome, both for the petitioner and the regulatory agency. Since there is no evidence that food from cloned animals poses any increased health risk to the consumer, it could be concluded that food from cloned animals should be approved for consumption. However, the paucity of evidence in the literature on this topic makes it impossible to provide scientific evidence to support this position.

In view of the challenges discussed about the application of profiling methods to evaluate unintended changes in compositional profile, coupled with the lack of data in the scientific literature, other methods to evaluate such changes are those wherein the analysis is targeted for a specific macronutrient, micronutrient, or metabolite. The methods for the nutrients listed in Box 1a and 1b have been developed, validated, and are sufficient for the compositional assessment. Clearly, this approach is dependent upon an approach wherein a number of targeted compounds are assayed. Determining the list of compounds to assay is important.

Since many constituents of food are biologically degraded by heat processing or cooking and the digestive process, it is expected that this would render them biologically inactive. There are notable exceptions, specifically those proteins that induce an allergic response. For the vast majority of other molecules, biological inactivation via cooking or digestion provide another means to assure that any unintended change in a food component are not associated with increased health risks to the consumer.

On October 31, 2003, FDA released a preliminary report in which it was concluded that "Edible products from normal, healthy clones or their progeny do not appear to pose increased food consumption risks relative to conventional animals" (FDA, 2003). At present, there is no supportive evidence for increased risk to consumers of animal products from cloned animals, with the exception that it would not be prudent to allow animals that are genetically engineered to produce pharmaceuticals or other biologics, such as silk, to enter the food chain.

Other Considerations

The committee was asked to present a position about whether transgenic animals classified as "no-takes" should be permitted to enter the food chain. No-takes are animals in which the transgene is not incorporated into the genome. Implicit to this is the use of molecular techniques, including southern blotting, noncoding gene integration tags to detect integration of a gene, and real-time polymerase chain reaction to detect the presence of the transgene and the number of copies integrated into the genome. If the transgene DNA is not detected, the animal is considered a no-take. A list of tissues to sample, which will assure the transgene is not expressed in one tissue, but is expressed in another, must be established. If animals are determined to be no-takes, there is no scientifically based rationale to exclude them from entering the food chain.

Animal Identification

A challenge to regulatory oversight of cloned and transgenic animals is development and implementation of effective programs to monitor the presence of these animals. The challenges this poses for transgenic animals have been reviewed by Howard and colleagues (2001), and an important issue to be resolved is whether cloned animals should be differentiated from noncloned animals at the point of entry of animals into the food system. Currently, there is no analytical method available to differentiate cloned from noncloned animals. However, should subsequent scientific review establish an increased risk to human health associated with the consumption of food products from cloned animals, it will be necessary to distinguish cloned animals from noncloned animals prior to entry into the food chain.

It is envisioned that the greatest likelihood of increased risk may arise from cloned, transgenic animals in which the genetic modification, as the result of transgenesis, has been made for the production of biomaterials or pharmaceuticals. In this case, methodological approaches are available to identify the transgene. This will be important in order to differentiate cloned, transgenic animals developed for food production purposes from cloned, transgenic animals developed for the production of biomaterials, pharmaceuticals, and other non-food purposes.

Due to the possible need to differentiate cloned, transgenic animals, a national system for animal identity and identity preservation is required. This system must be implemented at the point of slaughter or processing to rapidly and inexpensively identify the presence of cloned, transgenic animals or products derived from these animals. While the question of animal identification is beyond the scope of this report, it will likely be an important component of future oversight processes developed to monitor the entry of cloned, transgenic animals into the food system.

CONCLUSIONS

1. Profiling techniques are appropriate for establishing compositional differences between cloned and noncloned animals.

2. Profiling methods and their interpretation are not sufficiently developed to allow direct assessment of potential health effects associated with most unintended compositional changes.

3. There is no scientific basis to exclude animals deemed to be "no-takes" from entering the food chain.

4. There is a need to improve our ability to detect and assess the health

consequences of unintended changes in GM foods, such as better tools for toxicology assessment and a more robust knowledge base regarding which components impact health.

5. Given the possibility that foods with unintended changes could enter the marketplace, there is a need to enhance our capacity for postmarket surveillance of exposure and effects.

RECOMMENDATIONS

1. Targeted analysis of selected nutrients in animal products should be performed on a case-by-case basis.

2. Standardized sampling methodology, validation, and performance-based analyses should be performed for targeted analyses.

3. Standardized methodology, validation, and performance-based analyses should be performed for profiling analyses.

4. Publicly available compound identification databases should be developed that contain information such as mass spectra and nuclear magnetic resonance spectra where appropriate.

   a. Publicly accessible databases should also be developed for compound profiles of animal-derived foods (see also Research Recommendations in Chapter 7 of the main report).

5. Animal identity and identity preservation systems should be improved for tracking animals and animal products through the food chain.

REFERENCES

Bauman DE. 1992. Bovine somatotropin: Review of an emerging animal technology. *J Dairy Sci* 75:3432–3451.

Bauman DE. 1999. Bovine somatotropin and lactation: From basic science to commercial application. *Domest Anim Endocrinol* 17:101–116.

Brophy B, Smolenski G, Wheeler T, Wells D, L'Huillier P, Laible G. 2003. Cloned transgenic cattle produce milk with higher levels of ß-casein and k-casein. *Nat Biotechnol* 21:157–162.

CAST (Council for Agricultural Science and Technology). 2001. *Evaluation of the U.S. Regulatory Process for Crops Developed Through Biotechnology*. Issue 19. Washington, DC: CAST.

CAST. 2003. *Biotechnology in Animal Agriculture: An Overview*. Issue 23. Washington, DC: CAST.

Chilliard Y. 1989. Long-term effects of recombinant bovine somatotropin (rBST) on dairy cow performances: A review. In: Sejrsen K, Vestergaard M, Neimann-Sorensen A, eds. *Use of Somatotropin in Livestock Production*. New York: Elsevier Applied Science. Pp. 61–87.

Cummins JM. 2001. Cytoplasmic inheritance and its implications for animal biotechnology. *Theriogenology* 55:1381–1399.

De Sousa PA, King T, Harkness L, Young LE, Walker SK, Wilmut I. 2001. Evaluation of gestational deficiencies in cloned sheep fetuses and placentae. *Biol Reprod* 65:23–30.

Etherton TD, Bauman DE. 1998. The biology of somatotropin in growth and lactation of domestic animals. *Physiol Rev* 78:745–761.

Falk MC, Chassey BM, Harlander SK, Hoban TJ, McGloughlin N, Akhlaghi AR. 2002. Food biotechnology: Benefits and concerns. *J Nutr* 132:1384–1390.

FDA (U.S. Food and Drug Administration). 2003. Animal Cloning: A Risk Assessment, Draft Executive Summary. Washington, DC: U.S. Food and Drug Administration. Online. Available at http://www.fda.gov/cvm/index/cloning/CLRAES.doc. Accessed January 4, 2004.

Hill JR, Roussel AJ, Cibelli JB, Edwards JF, Hooper NL, Miller MW, Thompson JA, Loonery CR, Westhusin ME, Robl JM, Stice SL. 1999. Clinical and pathological features of cloned transgenic calves and fetuses (13 case studies). *Theriogenology* 51:1451–1465.

Hill JR, Burghardt RC, Jones K, Long CR, Looney CR, Shin T, Spencer TE, Thompson JA, Winger QA, Westhusin ME. 2000. Evidence for placental abnormality as the major cause of mortality in first-trimester somatic cell cloned fetuses. *Biol Reprod* 63:1787–1794.

Hill JR, Schlafer DH, Fisher PJ, Davies CJ. 2002. Abnormal expression of trophoblast major histocompatibility complex class I antigens in cloned bovine pregnancies is associated with a pronounced endometrial lymphocytic response. *Biol Reprod* 67:55–63.

Howard TH, Homan EJ, Bremel RD. 2001. Transgenic livestock: Regulation and science in a changing environment. *J Anim Sci* 79:E1–E11.

Jaenisch R, Wilmut I. 2001. Developmental biology. Don't clone humans! *Science* 291:2552.

Juskevich JS, Guyer CG. 1990. Bovine growth hormone: Human food safety evaluation. *Science* 249:875–884.

Karatzas CN. 2003. Designer milk from transgenic clones. *Nat Biotechnol* 21:138–139.

Kouzarides T. 2002. Histone methylation in transcriptional control. *Curr Opin Genet Develop* 12:198–209.

Kuhholzer B, Prather RS. 2000. Advances in livestock nuclear transfer. *Proc Soc Exp Biol Med* 224:240–245.

Kuiper HA, Kok EJ, Engel KH. 2003. Exploitation of molecular profiling techniques for GM food safety assessment. *Curr Opin Biotechnol* 14:238–243.

Niemann H, Kues WA. 2000. Transgenic livestock: Premises and promises. *Anim Reprod Sci* 60–61:277–293.

NRC (National Research Council). 1994. *Metabolic Modifiers: Effects on the Nutrient Requirements of Food-Producing Animals.* Washington, DC: National Academy Press.

NRC. 2002. *Animal Biotechnology: Science-Based Concerns.* Washington, DC: The National Academies Press.

OECD (Organization for Economic Cooperation and Development). 2001. *Consensus Document on Compositional Considerations for New Varieties of Soybean: Key Food and Feed Nutrients and Anti-nutrients.* Online. Available at htt[://www.olis.oecd.org/olis/2001doc.nsf/43bb6130e5e86e5fc12569fa005d004c/cdb400c627da47a5c1256b17002f840d/$FILE/JT00117705.PDF. Accessed January 16, 2003.

Polejaeva IA, Chen SH, Vaught TD, Page RL, Mullins J, Ball S, Dai Y, Boone J, Walker S, Ayares DL, Colman A, Campbell KH. 2000. Cloned pigs produced by nuclear transfer from adult somatic cells. *Nature* 407:86–90.

Rideout WM, Eggan K, Jaenisch R. 2001. Nuclear cloning and epigenetic reprogramming of the genome. *Science* 293:1093–1098.

Santos F, Zakhartchenko V, Stojkovic M, Peters A, Jenuwein T, Wolf E, Reik W, Dean W. 2003. Epigenetic marking correlates with developmental potential in cloned bovine preimplantation embryos. *Current Biol* 13:1116–1121.

Schilter B, Constable A. 2002. Regulatory control of genetically modified (GM) foods: Likely developments. *Toxicol Lett* 127:341–349.

Solomon MB, Pursel VG, Paroczay EW, Bolt DJ. 1994. Lipid extraction of carcass tissue from transgenic pigs expressing a bovine growth hormone gene. *J Anim Sci* 72:1242–1246.

Tamashiro KLK, Wakayama T, Akutsu H, Yamazaki Y, Lachey JL, Wortman MD, Seely RJ, D'Alessio DA, Woods SC, Yanagamachi R, Sakai RR. 2002. Cloned mice have an obese phenotype not transmitted to their offspring. *Nat Med* 8:262–267.

Westhusin ME, Long CR, Shin T, Hill JR, Looney CR, Pryor JH, Piedrahita JA. 2001. Cloning to reproduce desired genotypes. *Theriogenology* 55:35–49.

Willadsen SM. 1989. Cloning of sheep and cow embryos. *Genome* 31:956–962.

Willadsen SM, Polge C. 1981. Attempts to produce monozygotic quadruplets in cattle by blastomeres separation. *Vet Rec* 108:211–213.

Wilmut I, Schnieke AE, McWhir J, Kind AJ, Campbell KH. 1997. Viable offspring derived from fetal and adult mammalian cells. *Nature* 385:810–813.

Wilmut I, Beaujean N, DeSousa PA, Dinnyes A, King TJ, Paterson LA, Wells DN, Young LE. 2002. Somatic cell nuclear transfer. *Nature* 419:583–587.

Xue F, Tian C, Du F, Kubota C, Taneja M, Dinnyes A, Dai Y, Levine H, Pereira LV, Yang X. 2002. Abberant patterns of X chromosome inactivation in bovine clones. *Nat Genet* 31:216–220.